Intellectual Property in Chemistry

Intellectual Property in Chemistry

A Guide to Applying for and Obtaining a Patent for Graduate Students and Postdoctoral Scholars

Nelson Durán
Leandro Carneiro Fonseca
Amedea B. Seabra

CRC Press
Taylor & Francis Group
Boca Raton London New York

CRC Press is an imprint of the
Taylor & Francis Group, an **informa** business

CRC Press
Taylor & Francis Group
6000 Broken Sound Parkway NW, Suite 300
Boca Raton, FL 33487-2742

© 2019 by Taylor & Francis Group, LLC
CRC Press is an imprint of Taylor & Francis Group, an Informa business

No claim to original U.S. Government works

Printed on acid-free paper

International Standard Book Number-13: 978-1-138-60082-9 (Paperback)
International Standard Book Number-13: 978-1-138-60083-6 (Hardback)

Library of Congress Cataloging-in-Publication Data

Names: Duran, Nelson, author. | Fonseca, Leandro Carneiro, author. | Seabra, Amedea B., author.
Title: Intellectual property in chemistry : a guide to applying for and obtaining a patent for graduate students and postdoctoral scholars / author, Nelson Durán, Leandro Carneiro Fonseca, Amedea B. Seabra.
Description: Boca Raton : Taylor & Francis, 2019. | Includes bibliographical references.
Identifiers: LCCN 2018034876 | ISBN 9781138600829 (pbk. : alk. paper) | ISBN 9781138600836 (hardback : alk. paper)
Subjects: LCSH: Patent laws and legislation. | Chemistry—Patents. | Intellectual property.
Classification: LCC K1505 .D86 2018 | DDC 346.04/86—dc23
LC record available at https://lccn.loc.gov/2018034876

Visit the Taylor & Francis Web site at
http://www.taylorandfrancis.com

and the CRC Press Web site at
http://www.crcpress.com

Contents

Authors

Nelson Durán is a professor of chemistry and the biochemistry at the University of Campinas—UNICAMP (Brazil). He received his PhD at University of Puerto Rico (USA). He was associate professor at the Universidad Católica de Valparaiso, Chile, and completed a visiting professorship at University of São Paulo, Brazil. In 1978, he joined the Chemistry Institute of UNICAMP (Brazil) working in Biological Chemistry and Biotechnology. His present research interests are nanobiotechnology in cosmetics and pharmaceuticals, in addition to metallic nanoparticles as antibiotics and anticancer carriers, and in carbon and silica nanocarriers. Actually, he is Invite Professor at the Institute of Biology, Urogenital, Carcinogenesis and Immunotherapy Laboratory, Department Genetics, Evolution and Bioagents, at UNICAMP, Campinas, SP, Brazil. He is the coordinator of the Brazilian Network on Nanotoxicology, a member of National Institute R&D&I in Functional Complex Materials (INOMAT; MCTI/CNPq), vice coordinator of Laboratory of Synthesis of Nanostrutures and Biosystem Interactions (NanoBioss; MCTI), and member of Brazilian-NanoReg-European Community, *in vivo* nanotoxicology. He has published 463 ISI articles (h factor: 52) and 392 non-ISI articles; book chapters; and patents in chemistry, biochemistry, biotechnology, and nanobiotechnology and four books in the nanotechnology area ((1) N. Durán, L.H.C. Mattoso, P.C. de Morais (Eds). *Nanotechnology: Introduction, Preparation and Characterization of Nanomaterials and Examples of Application* (in Portuguese), ArtLiber, Editora, São Paulo, Brazil, pp. 208 (2006); ISBN 8588098334, 9788588098336 (2006). (2) M. Rai and N. Durán (Eds). *Metal Nanoparticles in Microbiology*, Springer Verlag, Germany, pp. 330 (2011); ISBN 978-3-642-18312-6 (2011). (3) N. Durán, S.S. Guterres and O.L. Alves (Eds). *Nanotoxicology: Materials, methodologies, and assessments*, Springer, pp. 412 (2014); ISBN 978-1-4614-8992-4 (2014). (4) M. Rai, C. Ribeiro, L.H.C. Mattoso, N. Durán (Eds). *Nanotechnology in Food and Agriculture*, Springer-Verlag, Germany, pp. 343; ISBN 978-3-319-14024-7(2015)). He registered 65 patents.

http://lattes.cnpq.br/6191239140886028.

Leandro Carneiro Fonseca is a PhD student at University of Campinas (UNICAMP—Brazil) and a patent specialist. He is majored in chemistry at the Federal University of São Paulo (UNIFESP—Brazil; 2011) and received his master's degree at UNICAMP (2014). He holds academic and industrial experiences, as well as a background of technological innovation and technology transferring. He worked at Companhia Brasileira de Cartuchos (CBC—Brazil) as a chemical analyst for 1.5 years, gaining experience in control quality and strategic projects for optimization of production, training of team, and development of human resources. He worked at Innovation Agency Inova UNICAMP (official patent office of the said university) as an innovation analyst for one year and three months, holding experience in advanced prior art searches of patents and articles (Derwent, Orbit, Web of Science, INPI (Brazil), USPTO (USA), EPO (Europe) or Espacenet, SciFinder, Google Patents, and Google Scholar) and in elaborating technical advices of patentability analysis, patent protection strategy, and patent drafting. He works as patent consultant for intellectual property professionals and students. His present research interests are the development and characterization of nanomaterials and nanocomposites (carbon and silica nanocarriers) for biological applications and its interaction to biological systems, regarding the area of nanobiotechnology. He has five patents in the area of chemistry, one book chapter, and nine articles.

Amedea B. Seabra is a professor of chemistry at the Federal University of ABC—UFABC (Brazil). She received her PhD at the University of Campinas (UNICAMP; 2006). Thereafter, she worked as a research fellow at UNICAMP (2006–2008), and she took postdoctoral studies at the Chemistry and Biochemistry Department, Concordia University in Montreal, Canada (2008–2010). From 2011 to 2016, she was a professor at Federal University of São Paulo (UNIFESP). In 2016, she joined the Center for Natural and Human Sciences at Federal University of ABC (UFABC) working with Biomaterials and Nanotechnology. Her present research interests are the preparation, characterization, and evaluation of the cytotoxicity of nitric oxide-releasing nanomaterials for biomedical and agricultural applications. At the present moment, she published 83 articles (h factor: 26 according to the Web of Science, ISI), 16 book chapters, and 16 patents. Recently, in 2017, she was the single editor of a book titled *Nitric Oxide Donors: Novel Biomedical Applications and Perspectives* (Elsevier; ISBN: 9780128092750, pp. 364). In 2017, she was selected as emerging investigators from the Americas: New Talent: Americas by the Royal Society of Chemistry. In 2018, she received the award: 13th edition of the Mercosul Science and Technology Award as Senior Researcher – from the Brazilian Ministry of Science, Technology, Innovation and Communications.

Link to her CV Lattes: http://lattes.cnpq.br/4549950625515088.

chapter one

Introduction

1.1 General aspects

The intellectual property right (IPR) law defines intellectual creations entitled to protection, how to obtain (or lose) IPR, how to use and benefit from IPR, and how to enforce IPR and obtain compensation for infringements. All of these aspects are important to understand well to avoid infringing other people's IPR.

Very recently, an important seminar of Royal Society of Chemistry Law Group, named "Introduction to Intellectual Property for Researchers" was presented on May 2016 in London. This document provided an introduction to IPR for researchers in chemistry. The document was ideal for any researcher with a primary or no knowledge of patent and other IPRs. This type of seminar is extremely important for the comprehension of this area in which all are interested, namely, university, industry, and any civil member of our community.

The intellectual property (IP) involves several groups, such as industrial property, copyright, software, variety of plants or plant breed, and circuit technology, as shown in Figure 1.1.

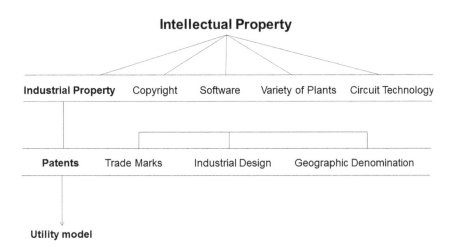

Figure 1.1 Modalities of intellectual property rights.

The copyright includes the following:

1. Copyright which covers (a) artistic, literary, and scientific works; (b) computer programs; and (c) scientific discoveries.
2. Related rights shall cover the interpretations and performances of performers, phonograms, and broadcasts.

Software: Although in the Brazilian IP computer programs are protected by copyright, these are analyzed by a specific legislation (Law No. 9.609, February 19, 1998) known as Software Law (www.planalto.gov.br/ccivil_03/Leis/L9609.htm, accessed March 5, 2017).

However, the IP protection related to computer software has been deeply discussed at various levels. In the European Union (EU), a Directive on the Patentability of Computer-implemented Inventions has also been debated for harmonizing the rendering of the national patentability demands for this kind of computer software-related inventions, including the business procedures carried out via the computer. The debate showed devious views among stakeholders in Europe. Besides, the Internet augmented complex issues concerning the performance of patents, as patent protection is fixed on a country-by-country basis, and the patent law only pays off within its own countries (www.wipo.int/sme/en/documents/software_patents_fulltext.html, accessed March 5, 2017).

Industrial property covers the following:

1. Patents which protect inventions in all the areas of human activity;
2. Trademarks and business names;
3. Industrial designs and models;
4. Geographical indications.

Plant variety covers the following: Cultivated, traditionally bred, medicinal and aromatic plants can also be protected by a different way of plant breeder's rights. In many cases, the countries offer special legal protection for the products of plant breeding applying the International Convention for the Protection of New Varieties of Plants (van Overwalle, 2006).

Topographies of integrated circuit covers the following: Integrated circuits (ICs), named also as "chips" or "micro-chips," are the electronic circuits where all the components, such as transistors, resistors, and diodes, have been mounted in a rigid order on the surface of a semiconductor material.

In the actual technology, ICs are important elements for a wide range of electrical tools or products, such as watches, TVs, home tools, cars, smart phones, and computers, among other digital devices. The innovative layout design of ICs is the basic for the manufacture of any smaller digital devices with different properties or functions. Albeit creating a new layout is normally the result of a huge investment, either in financial terms or in terms of the time required by highly qualified experts, reproducing such a scheme or layout design may cost only a fraction of the initial investment. With the view to avoid nonauthorized copying of this kind of designs and to provide incentives for applying investment in this area, the layout design, which is in this case topography of ICs, is protected by a *sui generis* IP system (www.wipo.int/patents/en/topics/integrated_circuits.html).

Since this book will be devoted to patent, the patent definition and the concepts involved with it will be explained in detail.

1.1.1   Patents

A patent is an official document issued by the government that describes an invention and furnishes an adequate right to exclude other parties from utilizing the invention for commercial objectives. This right is granted by a national government, upon application and pursuance, in exchange for the entire disclosure of an invention. The disclosure is first a confidential disclosure to the Patent Office which, in Brazil, Canada, and currently in the United States, becomes a nonconfidential disclosure to the public 18 months later. This kind of patent grants to the applicant exclusivity and rights to use or sell the information claimed in the invention for a short period of time. It is worth to observe the distinction between inventor and owner: an inventor obtains the patent issued in his or her name; however, an inventor always will be an inventor. The inventor may then assign the property of the invention to someone else. In many countries, patents have a lifetime of 20 years from the date of early filing and payment of the prescribed annual fees, of 17–20 years depending on the different countries (CAGS, 2015).

We have to keep in mind that a patent is not a discovery but an invention. To be acceptable for a patent, an invention is necessary to fulfill the following three main criteria: (a) must be novel or new, (b) must have some utility such as functional and/or operative, and (c) must not be obvious to a person skilled in the area of the invention.

Then, it is clear that a patent is granted for the physical embodiment of an idea or also applied to a process that produces something marketable or real, in other words, commercially negotiable. Products, processes, manufactures, or novel and useful compositions of matter, as well as not only any new and useful upgrading of these elements but also new uses of a known compound, are subjected to be a patentable matter.

Nonpatentable issues include ideas, scientific principle, theorems, or some invention that is illegal or involving illicit purposes. Natural phenomena and laws of nature are not eligible for patent protection.

Another possibility for IP right to protect inventions is the utility model. This right is at hand in many national statutes. It is closely to the patent but usually has a shorter term (often 6–15 years) and less rigorous patentability requirements. The utility models can be depicted as second-class patents (www.wipo.int/sme/en/ip_business/utility_models/where. htm, accessed March 6, 2017).

The terms of patents and utility models are represented in Figure 1.2.

IP offices in Asia registered the highest number of applications for patents and utility models. Specifically, a combined share of 60% of all patent applications worldwide came from Asian offices. These data contrast to the shares received by offices in North America (22.9%), Europe (12.9%), and Latin America and Caribbean (2.4%). In the case of utility models, Asia office was more significant (94.1%) than Europe (5.2%), North America (0.0%), and Latin America and Caribbean (0.5%).

In 2014, China reported the highest number of patent applications received by any single IP office, maintaining this position since 2011. China received more applications than Japan along with the United States. Middle-income countries, such as Brazil and India, rank among the top 10 despite having received fewer applications in 2014 compared to 2013. From the top 10 Intellectual Priority offices, China's IP office (+12.5%) saw the highest annual growth in filings received in 2014. On contrary, the office of the Russian Federation had a decline of 10.3% (Figure 1.3).

The IP office of China followed the same profile with the largest number of utility model applications in 2014, considering for just over nine tenths of the world total. The offices of Germany (14,741) and the Russian

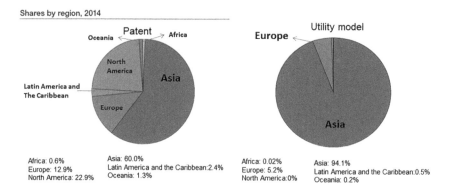

Figure 1.2 Pie charts present, for each intellectual property right, the distribution of intellectual property-filing activity across the world's six geographical regions. (Source: WIPO (Word Intellectual Property Organization), reproduced with modification.)

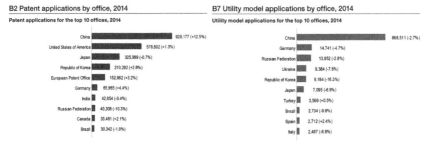

Source: WIPO Statistics Database, October 2015

Figure 1.3 Patent and utility model applications by office in the top 10 offices in 2014.

Federation (13,952) obtained similar numbers of applications, as close to the Republic of Korea and Ukraine with about 9,200 and 9,400, respectively. Recently, China saw a decrease in the number of applications filed at its office (Figure 1.3; http://www.wipo.int/edocs/pubdocs/en/wipo_pub_943_2015.pdf, 2016, accessed March 6, 2017; World Intellectual Property Organization (WIPO), 2016).

In Latin America, Brazil (61.0%), Argentina (16.3%), and Chile (8.1%) appear to be the most productive in patent during 1980–2015 (Figure 1.4).

Similar statistics were recently published by the Brazilian government (Instituto Nacional da Propriedade Industrial (INPI); www.inpi.gov.br/sobre/estatisticas; Jorge et al., 2017).

1.2 Patenting process through years

The first historical reference to an institution responsible for issuing and archiving patents was at 1679, with the creation of the General Board of Trade and Currency (GBTC) of Spain. This GBTC of Spain had the responsibility to increase economic growth. Invention rights in Spain were granted before 1679 by the King of Spain in the 15th and 16th centuries.

The first establishment of patent laws in the United States was in 1790, and the patent numbering started in 1836. During the International Exhibition of Inventions in Vienna (1873), foreign exhibitors refused to participate in this exhibition, because they were afraid that someone else would steal and commercially use all their ideas in different countries. From this confusing situation appear the needs to create a system to protect the IP. Ten years later (1883), the Paris Convention for the Protection of Industrial Property was created. This was the first international treaty created to help the people to have protection of their invention in a

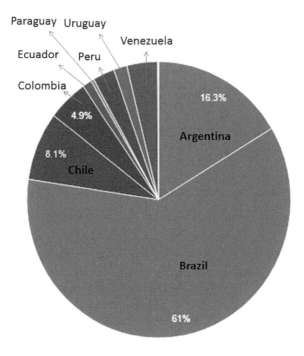

Figure 1.4 WIPO intellectual property in Latin America (Patent 1980–2015). (Source: WIPO (Word Intellectual Property Organization).)

particulate country in other country for their ideas and intellectual creation under the industrial property rights (IPR).

The U.S. Constitution gives to the Congress the power to decree laws relating to patents. Then, the Congress has decreed laws relating to patents until now. Important than in 1980, the Bayh–Dole Act gave universities title to the property of inventions resulting from research support by the federal government, since before that time, title ownership belonged to the government. In 1984, the Hatch–Waxman Act was approved. This act permitted generic drugs to enter the market. After 1984, the generic drug company was required only to demonstrate bio-equivalency of the generic drug. In 1886, the Berne Convention for the Protection of Literary and Artistic Works appears and in 1988 the European Union legislation, United Kingdom: Copyright, Designs and Patents Act. In Brazil, in 1996, the Law No. 9.279 of the Industrial Property Law was approved by the Brazilian government.

The American Inventors Protection Act (AIPA) of 1999 brought changes in the U.S. patent laws. Some of these changes were that a U.S. patent application was no longer a secret by the U.S. Patent and Trademark Office (PTO), but now is normally published 18 months after the application filing date. The patent protection is now adjusted to offset for delays provoked by the PTO during the examination. Besides the changes enacted by the U.S. Congress in the AIPA, initiatives at administrative level taken by the PTO at the end of 2000 were also largely changing the practice of patent law by giving way for online filing of a patent application and making accessible for online review of the PTO's internal data on the status of a patent application (McLeland, 2002; Waller, 2011).

chapter two

Patent profile

2.1   *Invention versus innovation*

As the definition of patent has been previously discussed in Section 1.1, now we will be discussing other details. In general, many people use erroneously INVENTION as the synonym of INNOVATION. Besides being incorrect, other aspects must be considered. The invention involves creating a new thing, whereas the innovation is related to a concept of using an idea or a process or a method. Although this difference is pervasive and, in many cases in dictionaries, is considered as synonymous, these two concepts are not interchangeable at all. The description of an invention is usually a "thing," whereas an innovation is usually an invention that causes change in behavior or interactions and finally giving a product or a process.

Many companies claim to be a leader in innovation, showing a patent portfolio as evidence. In some sense, yes, the patents are evidence of inventions, documenting this invention through a legal process. The utility of all the inventions of the company is not proven, so the inventions are not innovations. It is known that many patents have no practical use or have not influenced any products in the industry. Therefore, such patents without any use or applications are not innovation (Walker, 2015).

2.2   *The importance of intellectual property protection*

Patents can protect an existing business or a new one. Companies with patent are able to license it and gain royalties, and the owner will exclude any emerging technology that is new and on the cutting edge of science. Waller (2011) made an interesting assumption related to drug developmental costs since 1992. In 2007, the number of USD$2.8 billion represents the Exubera® failure by Pfizer®. Numbers are extremely variable depending on who is carrying out the analysis. Values range from USD$521 million or USD$868 million or USD$2.2 billion up to USD$2.9 billion (Millman, 2014). Anyways, Waller (2011) supposed that the value from Pfizer was reasonably correct. This company as any other should have protection from payoff copies of their products, and this is also an extra cost. Remember that the patent covers 20 years of protection of exclusivity of the product, selling, or other activity. In many cases, patents can be licensed to produce a revenue stream, and this, in general, represents an important profit to the company. But if all of these costs are not recovered, the company will fail after these 20 years of development. So, intellectual property (IP) is a mechanism to get hold of both the costs in bringing a successful product to the market and finally not bringing the unsuccessful products to the market.

Patents are significant corporate ownerships, and the cost to maintain the IP can be one of the most costly components for many products. Besides the above discussion of the importance of IP, the dark parts are the cost for these procedures. A good example was discussed by WIPO (2016): A system called MPEG-2 (a standard for the generic <u>coding of moving pictures</u> and also <u>ISO/IEC 13818 MPEG-2 at the ISO store</u>. It also describes a combination of <u>video</u> and <u>audio data compression</u> methods, which enable storage and transmission of movies using nowadays-available storage media and transmission bandwidth) is a technical standard for various consumer products dealing with video technology. The MPEG-2 licensing fee per DVD player costs USD\$2.50, and the DVD manufactures were agreed to pay with the view to be compatible with the MPEG-2 standard. Then, patent holder's group separately license their patents related to DVD technology with fees collectively paid of USD\$8.50. Therefore, the IP license fees related to DVD players reached the value of USD\$11.00. Thus, for a DVD player, the final value was USD\$44.00, about one quarter of its price related to IP.

There is one important aspect to consider. Although the company can first start with market advantages, there is the possibility that a competitor may also have learned about the product successfully. If we consider that at least one competitor can learn to make the product more efficiently and cheaply than the original manufacturer, this would be a major problem for the original manufacturer. But if at least the first company in the market has important IP rights, it may finally see its revenue decline as larger, and better competitors enter the market. In this respect, if the company can exploit its IP rights, the company may completely prevent the ability of other manufacturers to produce the product, but the company may also have licensing income that represents an important fraction of what its own benefits would be for selling the product (WIPO, 2007a,b).

2.3   *Patents as sources of advanced technologies*

There are two important categories in patents such as structured and unstructured items. The first one is coherent in semantics and formats across patents. The items in this category are patent number, filing date, inventors, assignees, and priority data. Meanwhile, the unstructured one is the text with the content with different styles and important aspects, such as descriptions and claims.

The description section (or specification) includes details on the previous patent filings and prior art such as a revision of scientific literature and a summary followed by itemized background to the claimed invention. This section will typically include examples that could be actual worked examples or paper examples. The claim section of this kind of document is normally accounted as the most important part of the document as it tells us what the applicant is actually claiming as an invention. The information that is claimed in a patent application must be supported by the description.

Researchers are in agreement that the patent gives a font of information that the company can use to earn a competitive advantage (Shih et al., 2010). Companies can use all of this information to follow technological developments; identify new trends in the competitor industry; identify new competitors through their activities and plans in the research and development activities, possible joint venture potential partners, and found opportunities for product licensing; and, finally, identify competence and potential collaborators (Kehoe and Xiao, 2001).

Besides all of these aspects, the patent information is extremely important to enhance the quality of new patents, know the actual situation of the business ambience, and identify alternative technologies (Barroso et al., 2009; Oubrich and Barzi, 2014). In research and engineering activities, the use of patent documents can have the following significances: to avoid expensive duplication of research work, to initiate research from a higher or new level of knowledge, to encounter a new problem through old strategies to generate new ideas, to recognize the extension of patent protection related to a particular area of technology, and to know the technical and commercial trends in any country of interest or certain areas of technology (Jansson, 2017).

chapter three

How to use databases in patent search

The use of patent database is crucial for the monitoring of published patents on a particular subject of interest. From the technological and innovation point of view, the strategic monitoring of published patents related to a new idea of product or a strategic analysis of important countries in which a certain patent was filed is important for academic research groups and research and development centers of excellence. In this context, the knowledge of patent search in databases provides information about inventions regarding a certain area. Basic searches are presented as quick and efficient strategies for the "qualitative" monitoring of patents, and in this panorama, databases such as Google Patents, National Institute of Industrial Property (INPI) of Brazil, Web of Science—Derwent, SciFinder, United States Patent and Trademark Office (USPTO), Espacenet, and Free Patents Online (FPO) are as commonly used and are presented in the following topics.

For the use of the presented platforms, it is necessary to comprehend the Boolean operators, those applied to the combination, and the refinement of keywords. The Boolean operators such as **AND, OR**, and **NOT** are explained below.

AND operator: This is used to search for documents that contain at least two keywords not necessarily close to each other, that is, they are presented in different or in the same fields from the patent document.

As an example, the hypothetical search (*fluorescence* **AND** *molecule*) will present documents in which both words are present and can be found in the following possible configurations:

1. The word <u>*fluorescent*</u> can be found, for example, in the *Field of the Invention* in the expression "<u>*fluorescent*</u> <u>*probe*</u>" and <u>*molecule*</u> may be found in the *Claims* in the expression "<u>*organic*</u> <u>*molecule*</u>." In this situation, both chosen words are so far from each other and are present in two different phrases expressing different ideas;
2. Both chosen words may be close to each other as a hypothetical expression "<u>*Fluorescent*</u> <u>*aromatic*</u> <u>*molecule*</u>" in any fields from the patent (Field of the Invention, Claims, Background of the Invention, Brief Description of the Invention, etc.);
3. The configurations 1 and 2 are simultaneously present.

OR operator: This is used to search at least two words usually designated as synonyms. In this case, the found documents may contain one, two, some, or all requested words. As an example, the search (*mixer* **OR** *mix* **OR** *composite*) finds documents for at least one of the words requested, in any possible combination and order.

NOT operator: This is used to subtract a set A of patents from a set M of patents, resulting in a set B where B = M – A; this operator may also be used to search for documents containing at least one word of interest and disregard documents containing another set of at least one chosen word. As an example, when searching for (*mixer* **OR** *mix* **OR** *composite*) **NOT** (*nanocomposite* **OR** *nanomaterial*), the found documents may comprise the words *mixer, mix,* or *composite* and will not necessarily cover the words *nanocomposite* or *nanomaterial*.

3.1   *Google Patents*

The access to the Google Patents database can be done through https://patents.google.com. This platform is useful when we already know a specific patent we want to analyze or when we want to qualitatively search for documents through the use of general keywords.

The Google Patents home page displays the search field where the number of a patent of interest can be typed. As an example, we searched the patent *US2011311029*, as shown in Figure 3.1.

Sequentially, the electronic page with the information of the patent of interest, as shown in Figure 3.2, will appear. The found patent is US2011311029A1 (patent filed). In the central part of the screen, the title

Figure 3.1 First electronic page of Google Patents. (Source: Google and the Google logo are registered trademarks of Google Inc., used with permission.)

of the patent (*Radiation Window, And A Method For Its Manufacturing*), the abstract, images, and international patent classifications (IPCs) can be checked—IPCs will be discussed in detail forward. In the right corner of the screen, important information of the patent is found and comprises, respectively, the following data (note: this it will be discussed in detail latter):

- Publication number: This is the observed number in the box (US2011311029A1) and refers to the published patent number (after the period of secrecy) found in the search;
- Legal status: This refers to the status (filed or granted) of the most recent patent in the patent family in which the found document is included;
- Application number: This refers to the patent number prior to publication;
- Other versions: This shows similar patents within the same patent family. In that case, US8494119B2 is included in the same patent family as US2011311029A1. This information also informs that the filed patent (A1) was subsequently granted (B2);
- Inventor: This indicates the inventors involved in the patent;
- Current assignee: This shows the current public or private institutions involved in the patent (this topic shows possible inclusion or exclusion of institutions in relation to the original assignees);
- Original assignee: This displays the original institutions (in the example in question the original and current institutions are the same);

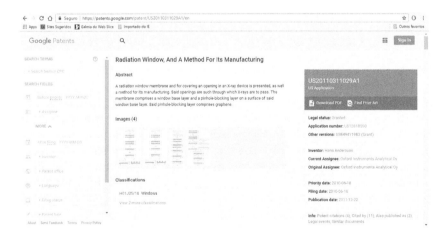

Figure 3.2 Detailed information of a patent through Google Patents. (Source: Google and the Google logo are registered trademarks of Google Inc., used with permission.)

- Priority date: This refers to the filing date of the oldest patent in the patent family;
- Filing date: This refers to the filing date of the found patent in the search;
- Publication date: This refers to the publication date of the found patent in the search;
- Info: This encompasses important information from the found patent, including citation data, publications, legal status, similar patent documents, and so on.

Further down the electronic page, the detailed information of the patent, including the full description of the invention at the left column and the claims at the right column, can be found. After the description of the document climbing down the electronic page, the following information is found: tables containing the cited patents from the found patent, patents that mention the found patent, table of the patent family of patents belonging to the found patent, table of documents (patents and/or scientific articles) similar to the found patent in the search, and table of description of the legal events related to the found patent.

Google Patents also offers patent search using keywords. So far, we have learned to search by the patent number and understand how to interpret the data in the way they are organized in an electronic page. In this way, by clicking the *magnifying glass icon* at the top of the screen, we will return to the home page, as shown in Figure 3.1, and a new search can be performed. Assuming that we want to find documents related to *silver nanoparticles* applied in *drug delivery processes*, we can type the following expression in the search field:

$$(\text{``}\underline{\textit{Silver nanoparticle}}\text{''}) \textbf{ AND } ((\text{``}\underline{\textit{drug delivery}}\text{''}) \textbf{ OR } (\text{``}\underline{\textit{drug loading}}\text{''}))$$

Prior to the request of the command search, it is necessary to understand the logic expressed in the above sets. First, the use of the double quotes in the expressions *silver nanoparticle*, *drug delivery*, and *drug loading* indicates that we want to find documents that contain the words exactly in the way they are written. In "*silver nanoparticle*", for example, the words *silver* and *nanoparticle* will be searched exactly side by side. The same reasoning applies to "*drug delivery*" and "*drug loading*." From the above expression, we have two sets A = ("*silver nanoparticle*") searching for words *silver* and *nanoparticle* necessarily together and B = (("*drug delivery*") **OR** ("*drug loading*")) designating the terms "*drug delivery*" and "*drug loading*" as synonyms. Finally, A **AND** B will search for the keyword sets A and B necessarily found in the same document, where A and B may be close to or distant from each other in the found patents due to the presence of the Boolean operator **AND**. The result of the search will be a reflection of patent documents with this logical keyword combination. After the search requesting, the obtained results will be displayed, as shown in Figure 3.3.

The central column of the screen will show the main found documents according to the combination of the requested keywords. In the central upper corner, the number of available documents is displayed in the information *about n results*, where $n = 195$ in that specific case. There is also a column in the left corner that provides the possibilities of refinement of results, explained below:

A. Search terms:

    I. **+ Search term or cooperative patent classification (CPC):** Other keywords can be written in this field and will be combined by applying the **AND** operator in relation to the current search. The refinement can also be accomplished by inserting the IPC. After adding keywords in this field, another field *+ Search term or CPC* will automatically appear to continue other possible refinements with keywords and/or CPC. Note that once entered, keywords can be edited or deleted.

B. Search fields:

    II. **Priority:** This refines documents according to priority dates prior to the entered date in this search field;

    III. **+ Assignee:** This refines documents by institutions according to the entered name in this search field;

    IV. **After filing:** This refines by documents containing filing dates subsequent to the inserted date in this search field;

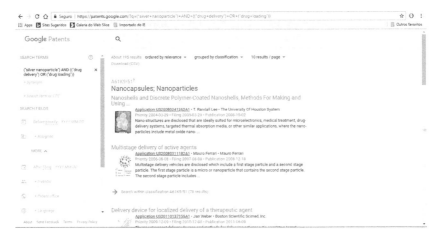

Figure 3.3 Electronic page of patent search results of Google Patents. (Source: Google and the Google logo are registered trademarks of Google Inc., used with permission.)

V. **+ Inventor:** This refines by the first or last name of at least one inventor inserted in this search field;

VI. **+ Patent office:** This refines by documents deposited by a particular entered patent office (e.g., US, EP, WO, CN, CA, KR, JP, etc.) in this search field. By clicking this field, patent office suggestions automatically will appear;

VII. **+ Language:** This refines by documents according to the entered language in this search field. When clicking this field, language suggestions are automatically visualized;

VIII. **Filing status:** This refines by patent filing status (granted or filed);

IX. **+ Patent type:** This refines by the type of document and may be chosen among groups of patents or drawings;

X. **+ Citing patents:** In this kind of refinement, the entered patent in this search field cites a group of patents that may be included among the documents available in the search we want to refine—it is important to know that at least one patent can be typed in this search field. More specifically, when we know that two patents X and Y cite a group of n documents possibly contained in the search of m patents that we want to refine (m contains n), the patent numbers of the documents X and Y can be entered in the + *Citing patents* field for refinement reasons. As a more elucidated example, the use of this kind of refinement may be important when we know the existence of patent A, which hypothetically cites 20 other patents containing different inventive concepts. Within these 20 cited patents, we wish to analyze only those that mention *nanoparticle issues* (we know that part of the 20 documents may mention *nanoparticle issues* that we want to discover, and the other part is related to another subject area that we do not need). One possible strategy is to initially enter *nanoparticle*-like terms and their possible synonyms in the search field + *Search term or CPC*, resulting in numerous possible documents in the state of the art. Next, we insert the patent number of the patent A in the + *Citing patents* field and request the refinement, resulting in eight documents (hypothetical number). We know that the resulting eight patents are found in the group of 20 patents cited by patent A. Therefore, the result allows us to conclude that from these 20 documents, 8 are related to nanoparticles;

XI. **+ CPC:** This refines by at least one international CPC inserted in this search field.

The Google Patents database is an excellent database for simple and quick patent searches. As noted, it can be used to find a known patent or search for a set of documents by simple combination of keywords and dynamic strategies for refinement of search.

3.2 Brazilian INPI

The INPI is the Brazilian patent office and can be accessed at www.inpi.gov.br, whose main page is shown in Figure 3.4. The site offers tools for simple and quick search of patents filed in Brazil.

The access to the INPI patent collection is performed through the link *Faça uma busca* (in Portuguese, *do a search*) which redirects to the page shown in Figure 3.5. This last page provides several search options within the context of industrial property.

By clicking *Patente* (in Portuguese, *patent*), we access the main INPI patent basic search database, as shown in Figure 3.6.

The electronic page displays some search fields related to Brazilian legal control. In this way, we will focus only on the fields related to the search for patents by the patent number and by combination of keywords following the basic and quick search strategy proposed in the Google Patents database. The patent search by the patent number (e.g., BR102016030134 or PI 0505217-3) is carried out in the field *Contenha o número do pedido* (in Portuguese, *contain patent number*). In this field, filed patents in the INPI until the end of December 2011 are searched under patent number in the old format PIXXXXXXXX. Patents filed from January 2012 meet the World Intellectual Property Organization (WIPO) standard formatting and are entered as BRXXXXXXXXXXXX. The search for patents by the combination of keywords (inserted necessarily in the Portuguese language) is carried out in the set of three search fields in the line that reads *Contenha* (in Portuguese, *contain it*).

Figure 3.4 First electronic page of the National Institute of Industrial Property of Brazil. (Source: National Institute of Industrial Property of Brazil.)

Figure 3.5 Intermediate electronic page of the National Institute of Industrial Property for patent search. (Source: National Institute of Industrial Property of Brazil.)

Figure 3.6 Electronic page of patent basic search of the National Institute of Industrial Property. (Source: National Institute of Industrial Property of Brazil.)

In the line *Contenha*, we call the three search fields as **left field, middle field**, and **right field**:

I. **Middle field (blank)**: It is a place where at least one keyword in Portuguese is inserted. Words are identified by backspace between them and are written without the use of any Boolean operator (**AND, OR,** and **NOT**);

II. **Left field (where *todas as palavras*, in Portuguese, *all words*, are initially read)**: By clicking *todas as palavras*, we have the following search options:

II-A. ***todas as palavras* (in Portuguese, *all words*)**: This searches for documents containing all the words inserted in the middle field and combine them using the **AND** operator (which will be implied between the inserted words in the middle field). For example, hypothetically inserting the words *grafeno prata* (in Portuguese, *graphene silver*) in the middle field and selecting the option *todas as palavras* in the left field, the search command will understand the configuration *grafeno* **AND** *prata*;

II-B. ***A expressão exata* (in Portuguese, *the exact phrase*)**: This searches for documents containing the words inserted in the middle field exactly in the way they are written, that is, they are searched side by side and exactly in the order they were inserted. As an example, by inserting the words *processo produção* (in Portuguese, *process production*) into the middle field and selecting the option *a expressão exata* in the left field, the requested search will understand the configuration "*processo produção*," that is, the word *processo* immediately on the left side of the word *obtenção*. The use of terms such as *de* or *da* (in Portuguese, *of the*), *com* (in Portuguese, *with*), *o* or *a* (in Portuguese, *the*), or any other possibility of prepositions, articles, and conjunctions is automatically omitted from the search, and thus, in the hypothetical possibility of insertion of the words *processo de obtenção* (in Portuguese, *process of obtaining*) in the middle field and selecting the option *a expressão exata* in the left field, the program will understand a similar search reasoning of the example above (similar to entering the words *processo obtenção* in the middle field);

II-C. ***Qualquer uma das palavras* (in Portuguese, *any of the words*)**: This searches for documents containing at least one of the entered words in the middle field by implicit application of the **OR** operator. For example, by hypothetically inserting the words *grafeno prata* similar to the example of topic **II-A** and selecting the option *qualquer uma das palavras* in the left field, the search command will understand the configuration *grafeno* **OR** *prata*;

II-D. ***A palavra aproximada* (in Portuguese, *the approximate word*)**: This searches for documents containing at least one of the words inserted in the middle field. Unlike the above options, the inserted words are searched by their derivation (suffix or prefix) or just the way they are written. For example, inserting the word *PEG* (reference to the acronym of the polymer polyethylene glycol) in the middle field and selecting the option *a palavra aproximada* in the left field, the search results will include documents containing the words <u>PEG</u> and their possible derivations such as <u>PEGUILADAS</u> (in Portuguese, <u>pegylated</u>), <u>NÃO-PEGUILADAS</u> (in Portuguese, <u>nonpegylated</u>), and <u>PEGUILAÇÃO</u> (in Portuguese, <u>pegylation</u>), among the several possibilities of variation of suffix and prefix of the radical <u>PEG</u>. Within the possibility of a certain search containing at least two keywords inserted in the middle field by executing the option *a palavra aproximada*, the program automatically implicitly inserts the **OR** operator among the terms;

III. **Right field (where *título*, in Portuguese, *title*, is first read)**: In this field, we select the location (in the documents from the search) in which we want the keywords to be present. By clicking the right field that is first read, *título*, we will find the following options:

    III-A. *Título* **(in Portuguese, *title*)**: The search results include documents whose requested keywords are present in the title of each found patent;

    III-B. *Resumo* **(in Portuguese, *abstract*)**: The search results comprise documents whose requested keywords are present in the abstract of each found patent;

    III-C. *Nome do depositante* **(in Portuguese, *name of the filer*)**: This option allows searching for the name of the assignee or person who filed the patent. The desired keyword is the name of the person or the name of the assignee;

    III-D. *Nome do inventor* **(in Portuguese, *name of the inventor*)**: This option allows searching for the name of the inventor involved. The desired keyword is the name of the inventor himself or herself. It is important to emphasize the significance of successive attempts by combining the first name of the inventor with his or her surnames in all possible combinations for the purposes of success in the search;

    III-E. *CPF/CNPJ do depositante* **(in Portuguese, *CPF/CNPJ of the filer*)**: It is possible to obtain patents through numbers referring to the Individual Taxpayer's Registry (CPF) of a person or through the number regarded to the National Registry of Legal Entities (CNPJ) of an assignee. In this case the keyword is a number.

Prior to the search request, the analyst has the option to choose the maximum number of patents per page to be displayed by selecting the respective quantity in the line that reads *Nº de Processos por página* (in Portuguese, *number of processes per page*). Defining the parameters of interest, the search is requested by clicking *pesquisar* (in Portuguese, *search*). As an initial example, we type the word <u>grafeno</u> in the middle field by

Figure 3.7 Electronic page of patent search results of National Institute of Industrial Property. (Source: National Institute of Industrial Property of Brazil.)

selecting the option *todas as palavras* in the left field and *título* in the right field and request the search. A list of patents similar to that shown in Figure 3.7 will be presented.

A list of found patents in a four-column table is shown in the first line:

Pedido/Depósito/Título/IPC
(Portuguese)
Patent number/Filing date/Patent title/IPC
(English)

Each patent can be accessed by clicking its patent number; all of them are displayed as links that redirect to detailed information on the patent application, as shown in Figure 3.8.

To facilitate the understanding of bibliographic data of the patent in Portuguese, Internationally agreed Numbers for the Identification of (bibliographic) Data (INID) codes are displayed. The most frequently encountered codes are as follows:

(21) Application number
(22) Filing date
(43) Publication date
(47) Granting date
(30) Priority details
(33) Country of priority date
(31) Application number in the priority country
(32) Priority date

Figure 3.8 Detailed information of a patent through National Institute of Industrial Property. (Source: National Institute of Industrial Property of Brazil.)

(51) IPC
(52) CPC
(54) Title
(57) Abstract
(71) Name of the person or assignee who filed the patent
(72) Name of the inventors
(74) Name of the attorney-in-fact
(85) Beginning of the national phase
(86) Application number and date of the international filing
(87) International publication number WO and date

The patent document is available in the lower central corner, below where it reads *documentos publicados* (in Portuguese, *published documents*). In this way, by clicking the cited link and typing the security code, the patent download in the PDF extension will be released.

3.3   Web of Science—Derwent basic

Web of Science is one of the most important platforms regarding the accessibility of patents and scientific articles for professionals in areas of Earth sciences. The significant and growing number of data added as well as the facilities of this search entity reflects its excellence and reference to scientists and engineers in the world. The access of patents is specifically done through the Derwent index, one of the sections of the Web of Science. In this context, the present topic is focused on the presentation of simple and quick search tools showing in a clear and dynamic way the facilities and differentials that this platform offers.

Derwent is accessed through the electronic page https://webofknowledge.com and by clicking the tab where it reads *Select a database* and selecting Derwent Innovations Index. We will automatically access the Derwent basic search page. The page features a single search field where they are entered as keywords and patent numbers when applicable. Beside the blank search field, a tab written *Topic* is found. By clicking it, we can choose the location where the keywords are able to be found within the patent document. These choices are explained below:

  I. **Topic:** It searches documents whose keywords are found in the title or abstract of the patent;
 II. **Title:** It searches documents whose keywords are found only in the title of the patent;
III. **Inventor:** It searches for documents by the name of at least one inventor. In this case, it is important to comprehend some important strategies to ensure the selection of a higher number of possible documents due to the countless manners the name of a certain inventor can be written in the patent and due to some punctuation and writing error commonly found. Especially for inventors having more than two surnames, we focused on a typical example of patent search for the inventor *Nelson Eduardo Durán Caballero* (one first name and three surnames): select the search option *inventor* in the gray tab and click the link *Select from index* which automatically appears. After typing the first name of the inventor (*Nelson*) in the blank search field below where it reads *Click on a letter or type a few letters from the beginning of the name to browse alphabetically by inventor,* click the button *Move to.* After that, an extensive list of possibilities of inventors named <u>Nelson</u> in alphabetical order of the initials of the possible surnames will be displayed. By clicking the buttons *next* or *previous*, it is possible to view the names *Nelson E, Nelson E D,* and *Nelson E C*, possibly in allusion to *Nelson Eduardo, Nelson Eduardo Durán,* and *Nelson Eduardo Caballero,* respectively. For all these possibilities, we click *add* to add these keywords in the search field below where it read *Transfer your selected inventor(s) below to the inventor field on the search page*—the keyword configuration initially will be viewed as NELSON E **OR** NELSON E D **OR** NELSON E C and will be saved as we look for other possibilities of combination names/surnames for *Nelson Eduardo Durán Caballero.* Other possible combinations are to find documents in which the inventor's name is started by the last name. In this context, we type the name <u>Durán</u> (or just <u>Duran</u>) in the search field *Click on a letter or type a few letters from the beginning of the name to browse alphabetically by inventor* and then we click *move to.*

Analyzing the new list using the *next* and *previous* buttons, we find *Duran C, Duran Caballero N E,* and *Durán Caballers N E*—maybe the Durán Caballero N E himself (?)—*Duran E, Duran N,* and *Duran N E C.* Adding all these names with the *add* button, the final configuration of keywords will be as follows: NELSON E **OR** NELSON E D **OR** NELSON E C **OR** DURAN C **OR** DURAN CABALLERO N E **OR** DURAN CABALLERS N E **OR** DURAN E **OR** DURAN N **OR** DURAN N E C (see it below where it reads *Transfer your selected inventor(s) below to the Inventor field on the search page*). Click the blue button *OK* and then a set of these keywords is automatically transferred to Derwent's main search field on the first electronic page. Then, we click *search*, and finally, we will find the results. It is important to note that there are different inventors whose first name and set of surname acronyms are the same and vice versa—as an example, it is possible that the DURAN N E C be, in fact, the *Durán Nelson Eduardo Caballero*; however, there are possibly several inventors called DURAN N. Regarding the *Duran Caballers N E,* we leave it to the reader to find out to whom it refers. One way to improve the accuracy of searches of this nature is to refine it by the name of the institution in which the inventor is bound—assignee refinement is explained further;

IV. **Patent number:** It searches for documents by patent number (to be written in place of the keyword in the search field);

V. **IPC:** It searches for documents by IPC (to be written in place of the keyword in the search field). In this option, the link *select from list* will appear below the gray tab to assist in the selection of IPCs through an attached list on the Derwent platform;

VI. **Derwent class code:** This is a differential feature that allows searching through the Derwent classification codes, which are grouped according to the large areas in which the inventions can be framed. More specifically, the classifications comprise the areas of chemistry (acronyms A–M), engineering (acronyms P and Q), and electrical/electronic (acronyms S–X). By clicking *Derwent class code*, the link *select from list* will appear below the gray tab to assist in choosing the Derwent classification codes through a list attached on the platform;

VII. **Derwent manual code:** This is another differential facility that allows searching for other types of codes, which are grouped according to the novel technical aspects of the invention and its applications. More specifically, the codes comprise the technical areas of polymers (A), pharmacological compounds (B), compounds with applications in agriculture (C), petroleum (H), and chemical engineering (J), among others. By clicking *Derwent manual code*, the link *select from list* will appear below the gray tab to assist in choosing the Derwent manual codes from a list attached on the platform;

VIII. Derwent prim. access. no.: It searches documents by Derwent primary accession number (PAN). Each patent document registered in Derwent holds a specific PAN, allowing this type of search when there is no knowledge about the patent number of a patent of interest or in the absence of any other information that favors a precise and direct search;

  IX. Assignee—name only: It searches documents by the name of the institutions involved;

   X. Assignee: It searches documents by the name of the involved institutions in an improved and complementary way in relation to the previous option. Derwent has specific codes for sets of institutions whose names are presented in many ways in numerous patent documents. For example, the company 3M holds patents in which its name can be observed as 3M of Brazil LTDA, 3M France SA, and 3M Cogent, Inc., among others. The code referring to 3M has been designated by Derwent as *MINN-C*, that is, by entering *MINN-C* in the search field, the final result will comprise documents whose 3M name will be displayed in at least three different manners (3M do Brasil LTDA, 3M France SA, 3M Cogent, Inc., or other possibilities). To find out the code of a particular institution, just select the option *Assignee* in the gray tab and then click the link *select from list*. On the next page, enter the name of the institution (3M, Basf, Unilever, Oxford, Harvard, etc.) in the search field below it reads *enter text to find a name or code containing or related to the text* and then click *find*. A list with the code (or codes) will appear and can be copied in the search field of the previous page. It is important to note that the search for institution names can be simultaneously performed using the *assignee-name only* and *assignee* options by adding the **OR** operator, whose application in the keywords of Derwent will be explained further.

The **AND, OR,** and **NOT** Boolean operators can be directly and randomly used between the keywords in the main basic search field of Derwent. Another way to use them is to add a new search field by clicking *+Add another field*. A gray tab will appear automatically to the left of the new added search field where Boolean operators are included. In this new situation, the new gray tab of operators **AND, OR,** and **NOT** combine the set of keywords entered in the first search field with the set of keywords of the second added search field. Still, a third search field can be included and combined with the other two search fields and so on. Note that for each added search field, the words can be randomly searched in any of the ten search options discussed above. After defining the combinations, the search can be requested by clicking *search*. Figure 3.9 illustrates the Derwent basic search home page and its many search possibilities.

   Important: In the above example, the keywords displayed in the sets (*silver nanoparticle*) and (*silver nanocomposite*) were inserted without

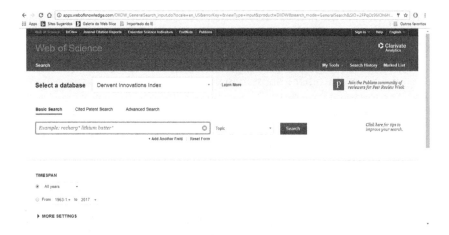

Figure 3.9 First electronic page of Derwent. (Source: Web of Science™ images are reproduced with permission from Clarivate Analytics © Clarivate Analytics 2018. Web of Science™, InCites™, Journal Citation Reports™, Essential Science Indicators™, Endnote™, Publons™, Clarivate Analytics™, and the logo of Clarivate Analytics are trademarks of Clarivate Analytics and its affiliated companies and are shown with permission. All rights reserved.)

inclusion of Boolean operators between each of them. In this situation, in the absence of any operator written in a particular search field, the program understands that among the keywords, there are implicit **AND** operators. For example, by typing the words (*electronic blue device*) and requesting the search, Derwent automatically understands the setting as (*electronic* **AND** *blue* **AND** *device*). By doing a search on the combination of keywords such as (*silver nanoparticle*)>**topic OR** (*silver nanocomposite*)>**topic**, we will have the result page as shown in Figure 3.10.

Exploring the above page, we have important information to explore, starting by observing the number of results found displayed in the upper left corner of the screen as well as a brief summary of the keyword combination in *You searched for*. There are, in the context of Derwent's basic search, refinement options shown in tabs in the column in the left corner of the screen. Among the various possibilities, the number of documents found can be better refined by subject areas, assignee names, inventors, IPC codes, Derwent class codes, and Derwent manual codes. To refine the results for a given option, simply click that option and select the items of interest that automatically appear after clicking the tab. After the choices, just click *refine*. Note that for each refinement tab, there is the link *more options/values …* link that displays additional refinement items. In succession, several refinement strategies using the left-hand column flaps can be performed until reasonable results are achieved for analysis.

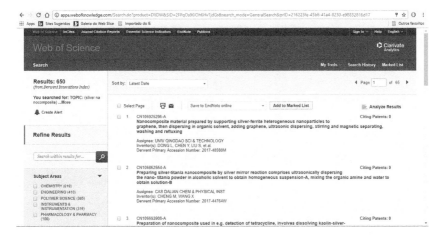

Figure 3.10 Electronic page of patent search results of Derwent. (Source: Web of Science™ images are reproduced with permission from Clarivate Analytics © Clarivate Analytics 2018. Web of Science™, InCites™, Journal Citation Reports™, Essential Science Indicators™, Endnote™, Publons™, Clarivate Analytics™, and the logo of Clarivate Analytics are trademarks of Clarivate Analytics and its affiliated companies and are shown with permission. All rights reserved.)

The central screen presents the main results covering the patents, which can be accessed by clicking the title of the document for viewing the abstract via Derwent or in the button *original* for direct access to the patent content. The abstract view opens the page shown in Figure 3.11. The information available in the splice includes the title and the patent number, the inventors and institutions involved, the Derwent PAN, abstract, IPCs, Derwent class codes, Derwent manual codes, and additional information in the lower corner such as filing and priority dates, country in which the patent was filed, and language, among other quick information for the patent analyst. In the context of this electronic page, the complete content of the patent can also be obtained by clicking the button *original* in the bottom corner of the screen, when there is one.

Derwent is a powerful patent search database, reflecting the numerous attachments, its constant updating, and the ease and possibilities of search strategies. Basic search on this platform is recommended for quick search situations without the need for a robust patentability analysis, since it includes, in the basic mode, features still limited when compared to the advanced mode as we will see later. However, it is important to note that Derwent offers a more accurate differential feature in document selection when compared to the Google Patents, INPI, SciFinder, USPTO, Espacenet, and FPO databases, mainly due to its greater availability of resources for the said purposes. We will see later how Derwent's

Figure 3.11 Detailed information of a patent through Derwent. (Source: Web of Science™ images are reproduced with permission from Clarivate Analytics © Clarivate Analytics 2018. Web of Science™, InCites™, Journal Citation Reports™, Essential Science Indicators™, Endnote™, Publons™, Clarivate Analytics™, and the logo of Clarivate Analytics are trademarks of Clarivate Analytics and its affiliated companies and are shown with permission. All rights reserved.)

advanced search enhances search accuracy and increases the chances of selecting documents that may not be found in a basic search.

3.4   SciFinder

SciFinder is the only searching platform that, unlike any presented database in this book, allows the search of patents (also for scientific articles) through molecular structures. More specifically, the program has a layout that allows the analyst to draw a molecule related to some invention and search for documents comprising drawings that are identical or slightly different from each other. In addition, this platform also covers a basic patent search system following the search logic studied further.

The SciFinder is accessed electronically at https://scifinder.cas.org starting at the page shown in Figure 3.12.

The SciFinder home page requests for the previously registered login where the user enters the username and password as shown above. By clicking *Sign In*, we will access the main search page of the program, as shown in Figure 3.13.

In the column at the left corner of the screen, we will focus on search options directly related to patent searches. Among the main ways of searching, we have the following:

  I. **Research topic:** This option presents a search field contained in the center of the screen where the keywords are entered. The latter

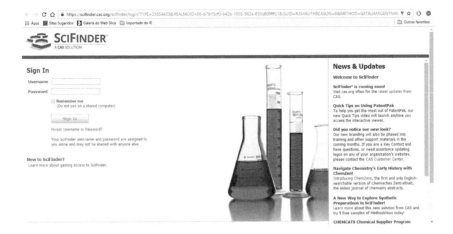

Figure 3.12 First electronic page of SciFinder. (Source: Image from SciFinder used with permission from CAS, a division of the American Chemical Society.)

Figure 3.13 Search electronic page of SciFinder. (Source: Image from SciFinder used with permission from CAS, a division of the American Chemical Society.)

are automatically recognized by spacing between them, in which the analyst can insert them in the presence or absence of Boolean operators **AND, OR,** and **NOT**. Regardless of whether or not to use these operators, SciFinder can perform at least two searches by automatically combining the entered keywords with **AND, OR** operators. In addition, the system also recognizes similar concepts or variations of the keywords as well as searches for documents containing the exact expression of the words written in the search field. The **NOT** operator, in turn, has to be entered manually when applicable, because the system does not recognize possible subtraction options. As an example, by typing *silver nanoparticle* in the search field and clicking the *Search* button, two types of combination of these words are proposed. In the first line, there are documents found by the exact phrase "*silver nanoparticle*," while in the second, the results cover articles and patents comprising similar keywords within the concept of *silver nanoparticle*. Select any or all of the resulted searches and click *Get References*. The subsequent electronic page will display the major articles and patents related to the subject, so we still need to refine them by patents (unlike the basic search database analyzed further, SciFinder searches for articles and patents on a single page of results, being necessary the refinement by only patent documents). In the left column, refinement possibilities including the one that selects by the type of document are available. For this purpose, click the *Refine* tab, select the *Document Type* and *Patent* items and click the *Refine* button below the column. The result page of a basic search by SciFinder is shown in Figure 3.14.

Note that through the left column of the result page, the search could still be analyzed by the name of authors (inventors in this case). Each patent document can be best viewed by clicking its respective link, as shown in Figure 3.15.

The electronic page of more detailed information of patent at SciFinder shows the patent title, application and publication numbers, legal status (filed or granted), priority, and filing dates. Unlike the programs studied in

Figure 3.14 Electronic page of patent search results of SciFinder. (Source: Image from SciFinder used with permission from CAS, a division of the American Chemical Society.)

Figure 3.15 Detailed information of a patent through SciFinder. (Source: Image from SciFinder used with permission from CAS, a division of the American Chemical Society.)

this book, SciFinder requires an extension of the program called PatentPak for viewing a patent document in the PDF version. In the absence of this program, we suggest copying the number of the patent of interest displayed on the screen of Figure 3.15 and inserting it into Derwent basic search by selecting the option to search by patent number. For the next search options in SciFinder, the search and refinement mechanisms are also similar. However, it is necessary to understand how to request the search commands through the other options described below. Access to the SciFinder main page is done by clicking the *Explore* tab and selecting the *Research Topic* option.

II. **Patent:** This searches for patent documents by patent number, assignee name, inventor name, and/or year of publication. Once the parameters have been selected, click the *Search* button;

III. **Chemical Structure:** This is the search mechanism that characterizes the differential feature of SciFinder since documents can be found through drawn molecular structures. In this context, the search result will encompass documents containing molecules specified by the analyst within the selected search conditions. By clicking *Chemical Structure*, some search options of drawing will be observed:

III-A. **Exact structure:** This searches for documents comprising the exact drawn molecular structure;

III-B. **Substructure:** This searches for documents comprising the exact or derived drawn molecular structure. For example, in drawing the *1,3,5-tribromobenzene* molecule, documents comprising structures such as *1,3,5-trichlorobenzene* or *1,2,3-tribrobenzene* could be found as they are derived molecules containing common aromatic ring and halogen groups;

III-C. **Similarity:** This searches for documents comprising the exact molecular structure, derived or similar to the one that was drawn. For example, when drawing the mentioned *1,3,5-tribromobenzene* molecule, documents comprising structures such as *2,4,6-tribromophenol·aziridine* (two interacting molecules constituting a possible compound) could be found in addition to the other mentioned possibilities in items II-A or II-B.

By performing the search for molecules click *Chemical Structure* and choose one of the three search options above. Selecting, for example, the substructure search option, click *Click to Edit* to open the structure editor window as shown in Figure 3.16.

In the example shown in Figure 3.16, the *1,3,5-tribromobenzene* structure was drawn using the editing tools contained in the left-hand column and the chemical element selection buttons, chemical bonding types, and predefined structures in the bottom corner center. Clicking the *OK* button, it will return to the previous page which displays the drawn molecule in the small frame where *Click to Edit* is read. Finally, by clicking *Search*, another search window will open and will display several sets of search results containing the exact molecule or derived structures. The sets are presented in the form of drawings of said molecules in which the analyst can select at least one and click *Get References* on the top of the screen. After this step, a new quick refinement window will appear and provide

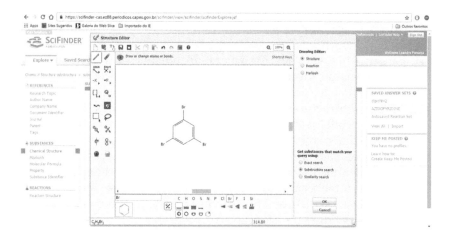

Figure 3.16 Structure editor window of SciFinder. (Source: Image from SciFinder used with permission from CAS, a division of the American Chemical Society.)

you with the possibility to search all sets of documents found without exception by selecting the *All substances* option. If the analyst wishes to search the selected set of documents and refine it, click *Selected substances* and select at least one of the refinement options at *Limit results to*. Among the several possibilities, the documents can be refined by preparation, process, crystalline structure, biological study, and several other options. Once the choices are made, click *Get* in the lower right corner, and the main results will appear. Note that the found documents correspond to articles and patents. To select only patents, simply refine the results by the type of document (patents) in a way similar to the search of the *Research topic* in item I;

IV. **Markush:** it is another search mechanism through molecular structure, which, in this case, refers to Markush structures. The search for Markush structures is similar to the search for substructures of item III-B. However, the difference in this case lies in the fact that the search is carried out taking into account the substructures also comprised in more complex structures and in the fact of maintaining the focus only on patents without the necessity of refinement by the type of document as observed in item I. By clicking *Click to Edit* to draw the molecule and selecting the *Markush* options in the upper right corner of the structure editor and *Substructures of more complex structures* in the lower right corner, click *OK* to return the previous window. Subsequently, the search is performed by clicking *Search*. The found results are consisted by only patent documents.

V. **Molecular formula:** This searches for documents by the molecular formula of a certain substance. By typing the molecular formula and clicking *Search*, the program will display result sets that can also be selected prior to the *Get references* command request. After the selection, the references are requested, and the quick refinement window will appear similar to the process in item III. After the respective selections, click *Get* for the search request, which will present related articles and patents. Thus, refining only by patent documents similar to the refinement process of item I, patents can be analyzed;

VI. **Property:** Interestingly, a search for scientific articles and patents can be performed knowing important physicochemical properties of a particular substance of interest. In this search option, the search is performed by the properties including boiling point, density, molar mass, vaporization enthalpy, and electrical conductivity, among others. More specifically, we have options of searching by experimental (by selecting the *Experimental* option) or theoretical values (selecting the *Predicted* option) and selecting one of the possible physicochemical properties that determine the search. For example, in the selection of *boiling point*

(°C) in *Predicted* and typing a hypothetical value of 100°C, the found documents will comprise articles and patents mentioning substances whose theoretical boiling point value is 100°C. Note that after clicking in *Search*, result sets will appear and should be selected according to the search strategy. Again, a quick refinement window will appear, and thus, the selected documents should be further refined by document type (patents) following the same reasoning as item I;

VII. **Substance identifier:** This searches documents by CAS number or a complete chemical name (common names, trade names, or acronyms are accepted). In this type of search, the results also cover scientific articles and patents, being necessary the refinement by document type (patents) according to item I.

As observed, SciFinder is presented as an important search tool mainly when organic molecules are objects of study since this program searches documents through a certain requested chemical structure. This feature is a differential one and complements the previous databases.

3.5   USA patents (USPTO)

The USPTO corresponds to the US National Patent Office and can be accessed at www.uspto.gov, as shown in Figure 3.17. This database owns the basic (quick search) and advanced searches. In this chapter, we will cover the basic search.

In the above electronic page, the access of USPTO basic patent search database is accomplished by clicking *Patents* in the upper left corner and clicking *Search for patents* below where *Application Process* is read (in the

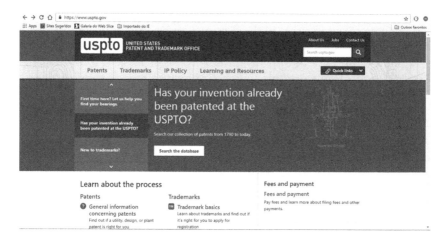

Figure 3.17 First electronic page of USPTO. (Source: United States Patent and Trademark Office, www.uspto.gov.)

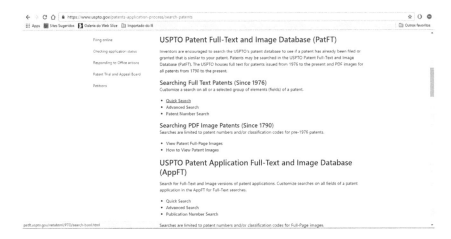

Figure 3.18 Intermediate electronic page of USPTO to access the basic search. (Source: United States Patent and Trademark Office, www.uspto.gov.)

top center of the screen). In the next page, the scroll bar is used to look for the basic search link below where *Search Full Text Patents (Since 1976)* is read, as shown in Figure 3.18.

By clicking the *Quick Search* link as noted in Figure 3.18, we will access the main electronic page of basic search of USPTO, where we will work with the keywords using the search tools offered by the program. By analyzing the structure of the site, two search fields named *Term 1* and *Term 2* are observed, both designed to insert keywords. There are two other fields identified as *Field 1* and *Field 2* that provide the search options. More specifically, the keywords typed in *Term 1* or *Term 2* will be searched according to the selected search option in *Field 1* and *Field 2*, respectively. Note that the chosen search option in *Field 1* can be the same or different from the one selected in *Field 2*. Advantageously, in the context of basic searches, USPTO offers several search options, that is, the chosen keywords can be found in the title or abstract of a patent, and the patent search can be performed by locations (city, state, or country); by name, city, state, or country of the involved assignee; by IPC; and by name, city, state, or country of the inventors, among several search options that can be chosen in *Field 1* and *Field 2*. Also, the typed keywords in *Term 1* can be combined with those typed in *Term 2* by inserting the Boolean operators **AND, OR**, or **ANDNOT** (similar to the **NOT** operator) included in a box where **AND** is first read between *Field 1* and *Field 2*. It is important to note that in the possibility of inserting more than one keyword in a certain field (*Term 1* or *Term 2*), the program implicitly inserts the **AND** operators between them. For example, by typing *lipid nanoparticle* in *Term 1*, the program comprehends the configuration *lipid* **AND** *nanoparticle*.

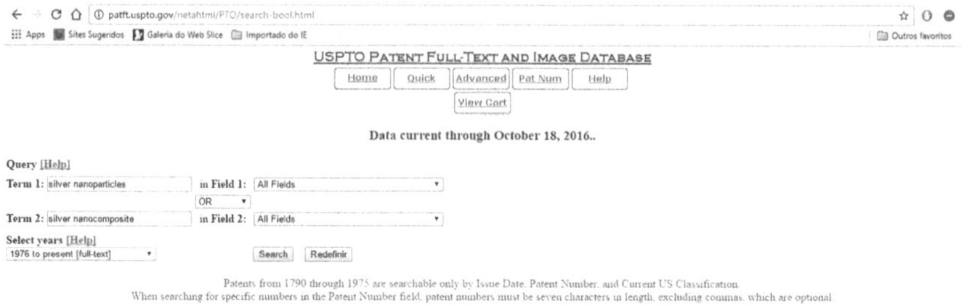

Figure 3.19 Electronic page of basic patent searches of USPTO. (Source: United States Patent and Trademark Office, www.uspto.gov.)

We start the USPTO basic search as shown in Figure 18: type *silver nanoparticles* in *Term 1* and select the *All Fields* search option in *Field 1*. Also, type *silver nanocomposite* in *Term 2* and select the search option *All Fields* in *Field 2*. The set of keywords in *Term 1/Field 1* is combined with the set of keywords in *Term 2/Field 2* set using the **OR** operator, and we request the search by clicking the *Search* button (Figure 3.19).

The electronic page of search results of USPTO is shown in Figure 3.20, which shows the organization of the documents by patent number and title.

The requested search configuration is noted where it reads, "*silver nanoparticles*" **OR** "*silver nanocomposite.*" Advantageously, USPTO provides a refinement field where it reads *Refine Search*, and therefore, it is important to understand its mechanism. Although the keywords were typed without quotation marks on the main USPTO search page, the program automatically inserts quotation marks to indicate different sets of

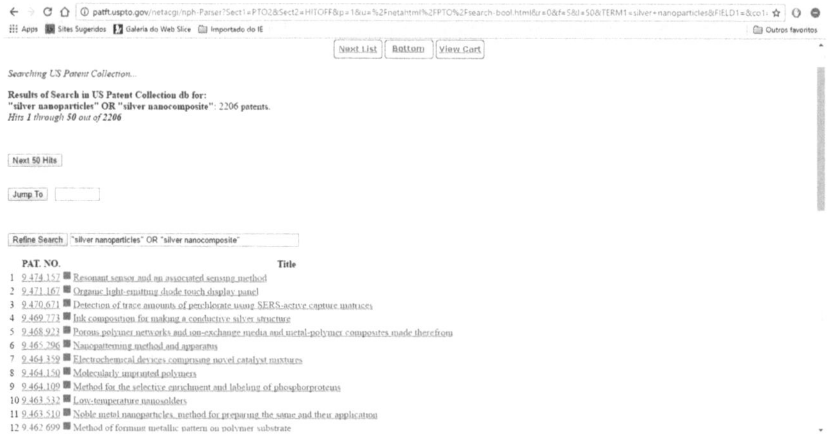

Figure 3.20 Electronic page of patent search results of USPTO. (Source: United States Patent and Trademark Office, www.uspto.gov.)

keywords, which in this case are the two sets *Term 1* and *Term 2*. Thus, in the refinement field (each set of keywords must be enclosed in quotation marks in the form *"keyword1 keyword2 keyword3."* Unlike the first search page, the refinement field allows the analyst to insert more than two sets of keywords and add other Boolean operators of choice. Look at the search configuration below:

"keyword1 keyword2" **OR** *"keyword3"*

We have two sets of keywords. If we want to insert another set of keywords, we would have the following:

"keyword1 keyword2" OR *"keyword3"* AND *"keyword4 keyword5 keyword6 keyword7"*

In this case, we have a third set of keywords and an **AND** operator randomly inserted—it is similar to mentioning that we have the supposed *Term 1*, *Term 2*, and *Term 3* (new inserted set). A practical example of this line of reasoning can be understood from the configuration below:

"Silver nanoparticles" **OR** *"silver nanocomposite"*

This is the beginning configuration of the search performed as shown in Figure 3.18, which could be modified as follows (among several possibilities):

"Silver nanoparticles" **OR** *"silver nanocomposite"* **OR** *"silver nanocomposites"*

In the above example, we consider the possibility of searching with the expression *silver nanocomposite* in the plural form (*silver nanocomposites*).

Another important aspect to be considered in the search refinement process above is the indication of the search option in which a certain set of keywords should be found (in the title, abstract, patent number, etc.). In this way, each search option has an associated code (CODE/) that must be inserted next to the set of keywords of interest. In this way, inserting the code in the set *"keyword1 keyword2"*, we have the following:

CODE/*"keyword1 keyword2"*

From the above configuration, it is understood that the words *keyword1* and *keyword2* should be contained in a given search option whose code is CODE/. Assuming the existence of three keyword sets such that the words of the first set should be searched in all fields of the patent, the second one only in the abstract (CODE/ = ABST/), and the third one only in the title (CODE/ = TTL/), we would have the following possible configuration:

"keyword1 keyword2" **OR** ABST/*"keyword3"* **AND** TTL/*"keyword5 keywor6 keyword7"*

Following the above reasoning, the search performed as observed in Figure 3.18 could be performed, for example, as follows:

"*Silver nanoparticles*" **OR** ABST/"*silver nanocomposite*"

The codes for each search option available in the USPTO are available at the end of this topic. After understanding the main refinement strategies possible in the USPTO, it makes it necessary to comprehend the process of visualization and interpretation of a particular patent using the own page of this database. Clicking the title of any found documents in the search of Figure 3.19 will allow the access of detailed information of the patent, as shown in Figure 3.21.

USPTO displays the full content of the patent of interest on the detailed information page, that is, in addition to showing important information such as title, abstract, filing date, inventor name, institution name, international classifications, and other crucial data for a quick analysis, the electronic page shows the detailing of the claims, field of the invention, and detailed description of the invention, among other topics. For the access of the document in PDF version, simply copy the patent number, insert it into Derwent, and get the original document.

Below it is presented the code list of search options available at USPTO for refinement purposes. Considering the range of specificities in the search options, each code could be better comprehended at http://patft.uspto.gov/netahtml/PTO/help/helpflds.htm#AFFF:

PN: Patent number
ISD: Issue date
TTL: Title

Figure 3.21 Detailed information of a patent through USPTO. (Source: United States Patent and Trademark Office, www.uspto.gov.)

ABST: Abstract
ACLM: Claims
SPEC: Description/specification
CCL: Current US classification
CPC: Current CPC
CPCL: Current CPC class
ICL: International classification
APN: Application serial number
APD: Application date
APT: Application type
GOVT: Government interest
FMID: Patent family ID
PARN: Parent case information
RLAP: Related US application data
RLFD: Related application filing date
PRIR: Foreign priority
PRAD: Priority filing date
PCT: PCT information
PTAD: PCT filing date
PT3D: PCT 371c124 date
PPPD: Prior published document date
REIS: Reissue data
RPAF: Reissued patent application filling date
AFFF: 130(b) Affirmation flag
AFFT: 130(b) Affirmation statement
IN: Inventor name
IC: Inventor city
IS: Inventor state
ICN: Inventor country
AANM: Applicant name
AACI: Applicant city
AAST: Applicant state
AACP: Applicant state
AACO: Applicant country
AAAT: Applicant type
LREP: Attorney or agent
AN: Assignee name
AC: Assignee city
AS: Assignee state
ACN: Assignee country
EXP: Primary examiner
EXA: Assistant examiner
REF: Referenced by
FREF: Foreign references

OREF: Other references
COFC: Certificate of correction
REEX: Reexamination certificate
PTAB: PTAB trial certificate
SEC: Supplemental exam certificate
ILRN: International registration number
ILRD: International registration date
ILPD: International registration publication date
ILFD: Hague international filing date

3.6   Espacenet

Espacenet is a worldwide patent database developed by the European Patent Office whose purpose is to quickly and objectively search for patents filed in European or international offices. One of the important features of this database is the translation tool in several languages, as we will see later. Espacenet is electronically accessed at https://worldwide.espacenet.com, as shown in Figure 3.22.

The main search field is available under Smart search, which is used to enter keywords or patent numbers. As an example of use, we type the keywords *silver nanoparticle* and click *Search*, accessing the page of results as observed in Figure 3.23.

The result page shows the main found documents briefly showing, for each patent, the name of inventors, name of the assignees, CPC and IPC, publication info, priority date, and title.

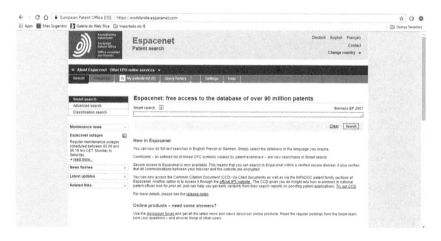

Figure 3.22 Main search page of Espacenet. (Source: European Patent Office, Espacenet.)

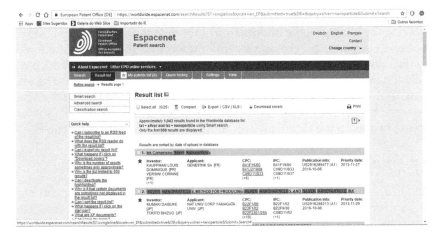

Figure 3.23 Electronic page of patent search results of Espacenet. (Source: European Patent Office, Espacenet.)

Clicking the title of any found documents found, the details can be accessed through the website. In this said next page, the abstract, the description of the patent, the claims, the original document available for download, and the bibliographic data can be found. Bibliographic data are the first information on the main page (Figure 3.24) after clicking the title of the patent on the result page.

In the upper left corner, a column containing more information from the found document is observed. By clicking *Description, Claims,* and *Original Document*, this information will be displayed. Espacenet offers a

Figure 3.24 Detailed information of a patent through Espacenet. (Source: European Patent Office, Espacenet.)

language translation tool. In the case of the electronic page of Figure 3.24 showing the bibliographic information of the patent of interest, the abstract could be translated by clicking the red button *patenttranslate* in the lower central corner. Prior to requesting this command, the language in which we want the translation to be performed is selected in *Select Language*. The text translation option is also available in *Description* and *Claims* topics following the same line of reasoning. To download the patent, simply click *Original Document* and on the *Download* link above the original document pages.

3.7   FPO

FPO is a search platform oriented to the search of US and European patents and could be used as a complement to the last studied databases. The electronic page is accessed at https://freepatentsonline.com and by clicking the *Search* link in the upper left corner that opens the page where we click *Quick Search*.

FPO database allows searching using the Boolean operators **AND**, **OR**, and **NOT**, and inserting quotation marks into phrases for searching according to the exact typed expression. By typing at least two keywords in the absence of typed Boolean operators, program comprehends all terms combined with the **AND** operator (implicit between words). For example, by typing *graphene oxide*, the search command is comprehended as being *graphene* **AND** *oxide*. In the possibility of typing "*graphene oxide*", program comprehends as being the exact expression *graphene oxide*, that is, both words side by side—note that in the use of quotation marks, there is no implicit **AND** operator.

It is interesting to note that FPO offers several search options that will be addressed below. Among the possibilities, documents can be found by patent number, keywords, inventor names, and assignee names, among others.

   I. **Number fields**
      I-A.   **Document number:** This searches for documents by patent number;
      I-B.   **Application number:** This searches documents by application number. In this case, the six-digit US patents may be inserted without the use of the US prefix or European patents with the prefix EP containing 11 digits;
   II. **Common fields**
      II-A.   **All:** This searches for documents in which the keywords are included in all fields of each patent (title, abstract, claim(s), and description/specification);
      II-B.   **Title:** This searches for documents in which the typed keywords are included in the title of each patent;

II-C. **Abstract:** This searches for documents in which the typed keywords are included in the abstract of each patent;

II-D. **Claim(s):** This searches for documents in which the typed keywords are included in the claim(s) of each patent;

II-E. **Description/specification:** This searches for documents in which the typed keywords are included in description or specification of each patent;

III. **Data fields**

III-A. **Filing date:** This searches documents establishing a period in which patents have been filed. In this field, the start and end dates that delimits the time can be selected;

III-B. **Publication date:** This searches documents establishing a period in which patents have been published. As in topic III-A, the start and end dates are selected to delineate the time of interest;

III-C. **Foreign priority:** This searches for documents by setting a priority date, that is, by choosing a specific day, month, and year, the program will search for documents according to the typed priority date;

IV. **Classification**

IV-A. **Current US classification:** This searches for documents by US patent classification;

IV-B. **International classification:** This searches for documents by IPC;

V. **Inventor fields**

V-A. **Inventor:** This searches for documents by inventor name. In this field, the search by name of the inventor is carried out following the format "last name first name";

V-B. **Country inventor:** This searches for documents by the inventor's country name. In this field, the country of choice is selected by typing its corresponding code according to an accessible list. For example, in the choice of resident inventors of Brazil, the BR code is used;

V-C. **Inventor state:** This searches for documents by the state name of the inventor in US territory. In this field, the state of choice is selected by typing its corresponding code as an accessible list. For example, in the choice of resident inventors of Alabama, the code AL is used;

V-D. **Inventor city:** This searches for documents by the name of the inventor's city by typing the name of the city from anywhere in the world;

VI. **Assignee fields**

VI-A. **Assignee:** This searches for documents by the name of the assignees in the patent. In the case of a company, its name is typed. In the case of a person, the name in the format "last name first name" is typed;

VI-B. **Assignee country:** This searches documents by the name of the institution's country. In this field, the country of choice is selected by typing its corresponding code according to an accessible list. For example, in the choice of resident inventors of Japan, the code JP is used;

VI-C. **Assignee state:** This searches for documents by the assignee's state name. In this field, the state of choice is selected by typing its corresponding code according to an accessible list. For example, in the choice of resident inventors of New York, the code NY is used;

VI-D. **Inventor city:** This searches documents by the name of the assignee's city by typing the name of the city from anywhere in the world;

VII. References

VII-A. **Domestic references:** This searches US or EP patent documents that cite US or European patents containing the typed patent number or application number in that field. By inserting, for example, a hypothetical patent USXXXXX or EPYYYYY, the search will return (a) US documents mentioning the USXXXXX patent if it is typed in the field or US documents mentioning the EPYYYYY patent if it is typed in the field or (b) EP documents mentioning the patent USXXXXX if it is typed in the field or EP documents mentioning the EPYYYYY patent if it is typed in the field;

VII-B. **Foreign references:** This searches US or EP patent documents that cite international patents containing the patent number or application number typed in that field. When typing, for example, a hypothetical patent JPXXXXX or BRYYYYY, the search will return (a) US documents quoting the JPXXXXX patent if it is typed in the field or US documents quoting the BRYYYYY patent if it is typed in the field or (b) EP documents mentioning the JPXXXXX patent if it is typed in the field or EP documents mentioning the BRYYYYY patent if it is typed in the field;

VII-C. **Other references:** This searches US or EP patent documents that cite patents containing word or phrase typed in the field. When entering, for example, *optical device* in that field, the search will return US or EP documents that quote patents in which those keywords are covered;

VIII. Legal/prosecution information

VIII-A. **Parent case information:** This searches for documents that are derived from the typed patent in this field as patent number or application number. Patents may be derived by processes of continuation, continuation in part, division, divisional, and so on;

VIII-B.   **Primary examiner:** This searches for patent documents by the name of the primary examiner. Words are typed in the format "last name first name";

VIII-C.   **Assistant examiner:** This searches for patent documents by the name of the assistant examiner. Words are typed in the format "last name first name";

VIII-D.   **Attorney or agent:** This searches for patent documents by the name of the attorney or agent. The words are typed in the format "last name first name" in the case of a person or the name of the company in the case of a firm.

Once the search strategy is established within the several possibilities observed above, it is necessary to choose in which groups we want the documents to be found. More specifically, the patents to be found may be within *US patents* or *US patent applications*, *EP documents, Japan abstracts, WIPO (PCT), German patents (beta)*, and/or *nonpatent literature*. These options can be freely selected in the bottom corner of the screen, where refinement options are also observed by filing period (patents filed in the *last 20 years* or *all years*), by *word stemming*, or by order (*chronological* or by *relevance*). The option of enabled *word stemming* (on) takes into account possible variants of any of the typed keywords in the eight search options above. By typing, for example, the word _metal_, the program can search for documents that cover possible derived words such as _metallic_, _metal_, _metallurgy_, and so on. In the possibility of *word stemming* off, the program considers the keywords exactly in the way they were typed in the above options. Finally, the search can be performed by clicking the *Search* button at the bottom of the screen.

    We have described the main databases for quick and simple patent searches in which the differential and complementary feature of each of them is clearly and objectively evidenced. In the subsequent chapter, therefore, the fixing exercises for learning and consolidation will be exposed, and later, the advanced patent searches in the main databases will be studied.

chapter four

Practical exercises in patent search

4.1   Exercises

1. Using the Google Patents database, you are asked to
   a. find the patent US20090088679;
   b. cite the inventor names of the found patent and its title;
   c. find what is the protection covered in the **first independent claim**.
2. Using the Google Patents database, you are asked to
   a. describe the protection covered in the **second independent claim** of a patent about supported polyoxometalates filed in 2006. One of the inventors of that document is Ryan M. Richards;
   b. cite the title, patent number, and patent assignee of the above found document.
3. Using the Google Patents database, you are asked to
   a. describe the title of a patent about mesoporous silica filed in 2008 by Virginia Tech Intellectual Properties;
   b. describe the patent number and inventor names.
4. Using the Instituto Nacional de Propriedade Industrial (INPI; Brazilian Patent Office) database, find
   a. the inventor names and patent number of the document about anticancer therapy with saponins (in Portuguese, *terapia anticancer com saponinas*). Tip: This patent was filed in Brazil in 2003 and its assignee is Panagin Pharmaceuticals, Inc.;
   b. the oldest patent of this patent family, its number, and filing date.
5. Using the INPI (Brazilian Patent Office) database, find
   a. the inventor names and patent number of a document about enzymatic inhibitor of gastric acids (in Portuguese, *inibidor de ácidos gástricos*) filed in 2006;
   b. the assignee name of this patent and the filing date of the oldest patent of this patent family.
6. Using the INPI (Brazilian Patent Office) database, find the patent PI0721626-2 and indicate the title and inventor names of this patent.
7. Using the INPI (Brazilian Patent Office) database,

    a. find the inventor names and the patent number of the document about spray with toothpaste (in Portuguese, *spray com creme dental*);

    b. indicate the codes of the two international patent classifications (IPCs) to which this patent is regarded to;

    c. through the site http://web2.wipo.int/classifications/ipc/ipcpub?notion=scheme&version=20170101&symbol=none&menulang=en&lang=en&viewmode=f&fipcpc=no&showdeleted=yes&indexes=no&headings=yes&notes=yes&direction=o2n&initial=A&cwid=none&tree=No&searchmode=smart, indicate the areas related to the two IPCs indicated in item b.

8. Using Derwent basic database, indicate the number of patents involving L'Oreal in 2017.

9. Using Derwent basic database, indicate the number of patents of researcher Lauro Tatsuo Kubota in 2017.

10. Using Derwent basic database, search for patent WO2008034207. Cite the title and assignee(s).

11. Tetrazene, of molecular formula N_4H_4, is an organic molecule used in explosive initiator mixtures as a sensitizer component. Its molecular structure is shown in Figure 4.1.

    Using the Markush structure editor from the SciFinder database, cite the title and assignee of the patent filed in 1992 citing the above structure. Important: use the exact structure search option.

12. In 1997, a patent was filed for the purpose of obtaining a pharmaceutical active compound whose molecular structure is shown in Figure 4.2.

    Using the Markush structure editor from the SciFinder database, cite the name of this pharmaceutical active compound, the title of the patent, and the assignee.

13. In 1858, the first diazo compound was produced by the German chemist Peter Griess. Its molecular structure is shown in Figure 4.3.

    One of the oldest patents related to the production of this compound was filed in 1923. Using the Markush structure editor from the SciFinder database, mention the name of this compound and the patent number and the intrinsic physical–chemical properties of this compound.

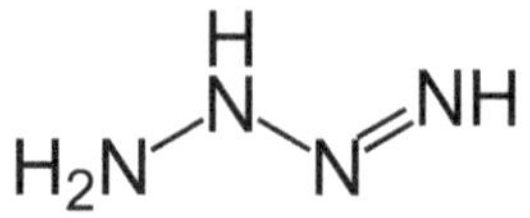

Figure 4.1 Tetrazene chemical structure.

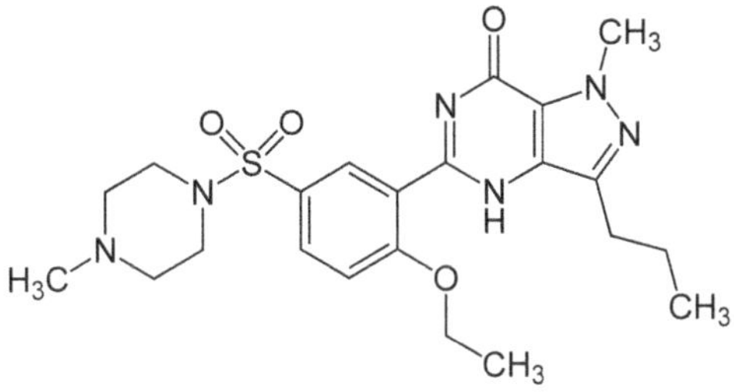

Figure 4.2 Chemical structure of a pharmaceutical active compound.

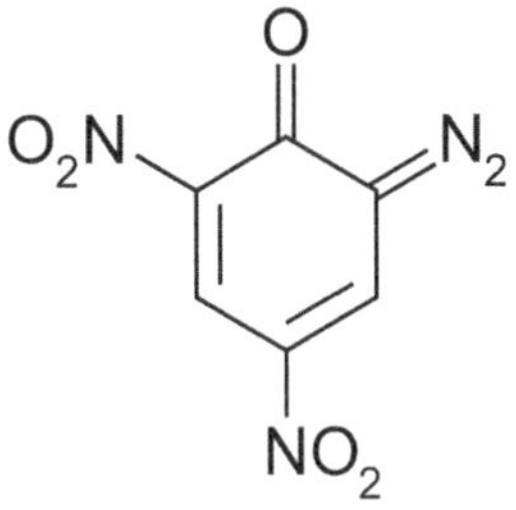

Figure 4.3 Diazo chemical structure.

14. In the context of intellectual property, it is very common to know important companies holding several patents in the international scene as IBM, Microsoft, Samsung, Google, and Apple, among others. Interestingly, there are famous celebrities in the world who also hold interesting inventions. One of them is an artist born in August 29, 1958, in the city of Gary (United States), known for his pop music and his original dance style, making it a worldwide success.
   a. Using Google Web, find out who the famous artist is;
   b. Using the United States Patent and Trademark Office (USPTO) database, mention the title and patent number of this famous artist filed in 1993. Tip: For the search by inventor name, enter the words in the format *Last Name First Name Abbreviated Middle Name*.

15. An internationally famous director and producer holds a patent filed in the United States whose patent number is US7366979. Using the USPTO database, find out the name of this celebrity and the document title. Tip: In searches using the patent number in USPTO, only numbers are used.

16. Using the USPTO database and Google Web, discover the name of the scientific television program presented by the famous inventor of the patent titled *Remote control device with gyroscopic stabilization and directional control.*

17. Using the Espacenet database, find a document about graphene oxide filter and its toxicity and mention the title, patent number, and involved inventors. For search reasons, consider patents filed up to 2017.
18. Using the Espacenet database, find the patent US2013074735 and mention the title and involved inventors.
19. Using the Free Patents Online database, find a patent related to a photovoltaic hybrid electrolysis cell filed in Europe by inventors from Berlin in the last 20 years and mention the title, patent number, and involved inventors.
20. Van Halen is a famous American hard rock band formed in Pasadena, California, in 1972. The band's guitarist, Eddie Van Halen (Edward) has worldwide fans. Despite known as a talented guitarist, few people know that Eddie is also an inventor and holder of more than one patent, all of them filed in the United States. Using the Free Patents online database, we ask
    a. the number of patents filed by Eddie in the United States;
    b. the title and patent number of the oldest filed patent;
    c. the title and patent number of the patent published in 2004.
21. Using the Free Patents Online database, find a patent about nano gaps and graphene filed by Oxford University in 2015 in the United States and mention the title, patent number, and name of the involved inventors.

4.2   Expected answers

1a. For the resolution of this item, the patent number US20090088679 is entered in the Google Patents main search field.
1b. The inventors of the patent are Kris Wood, Nicole Zacharia, Paula Hammond Cunningham, and Dean Delongchamp.
1c. The first independent claim protects the *composite thin film comprising the plurality of alternating layers of net positive and negative charge, wherein at least a portion of the positive layers, the negative layers, or both comprises a polyelectrolyte; wherein the layers are stable with respect to delamination at a predetermined voltage; and wherein the thin film is stable at a predetermined second voltage.*
2a. One of the possible ways to find this document is by typing *supported polyoxometalates* into Google Patents main search field. Numerous documents will appear in the search result which can be refined by typing the inventor name into the + *Inventor field* (Ryan M. Richards). After this, there are two options as follows: you look for the patent filed in 2006 through the result pages or enter the period in the *After filing* and *Before priority* fields. In this context, enter *2005-12-31* and *2007-01-01* in the *After filing* and *Before priority* fields, respectively. This search command will cover all patents from 2006 since

documents filed after December 31, 2005, and published before January 1, 2007, will be found. The second independent claim of the only found patent is the claim 13, whose disclosure is shown on the detailed patent description page of Google Patents.

2b. The title of the patent is *Supported polyoxometalates and process for their preparation*. The patent number and assignee are, respectively, US7417008 and ExxonMobil Chemical Patents, Inc.

3a. For the resolution of this exercise, we type *mesoporous silica* in Google Patents main search field. Countless documents will appear. Next, enter *Virginia Tech Intellectual Properties* in the *+Assignee* field, significantly reducing the number of results. For the selection of the document filed in 2008 by said assignee, the simplest way is to look at the result page. More precisely, the document can be found by typing *2007-12-31* and *2009-01-01* in the *After filing* and *Before filing* fields, resulting in only two documents. The one whose filing date is referring to the year 2008 is selected. The title of the patent is *Hybrid organic-inorganic gas separation membranes*.

3b. The patent number is US7938894. The inventors are Shigeo Ted Oyama, Yunfeng Gu, Joe D. Allison, Garry C. Gunter, and Scott A. Scholten.

4a. For the resolution of this exercise, the search field *Contenha* is used. Selecting the *Todas as palavras* and *Título* options in the left and right fields, respectively, type the word *saponinas* in the middle field and click *pesquisar*. There will be two filed patents in 2003 regarding concepts about saponins. Viewing both, select the one whose assignee is Panagin Pharmaceuticals, Inc. This said patent is filed as the patent number PI0311866-5. The inventors are M.J. Story and K.M. Kayne.

4b. The oldest patent in this patent family is 2,390,290, filed in Canada on November 6, 2002.

5a. For the resolution of this exercise, the search field *Contenha* was selected. Selecting the *Todas as palavras* and *Resumo* options in the left and right fields, respectively, enter the expression *inibidor gastricos* in the middle field and click *pesquisar*. The only patent filed in 2006 is shown in the results, whose patent number is PI0606333-0.

5b. The assignee name of this patent is Polyzen, Inc. The oldest patent of this patent family, which was filed in the United States, is the one whose patent number is 60/643,137, the filing date being December 1, 2005.

6. For the resolution of this exercise, we use the field *Contenha o Número do Pedido* and enter the number PI0721626-2. From the obtained result, the patent title is *Combinação farmacêutica sinergística para o tratamento de câncer*. The inventors of the patent are K. Joshi, M. Rathos, S. Sharma, and H. Khanwalkar.

7a. For the resolution of this exercise, the search field *Contenha* is used. Selecting the *Todas as palavras* and *Título* options in the left and right fields, respectively, enter the expression *spray creme dental* in the middle field and click *pesquisar*. The title of the found patent is *Escova de dente com creme dental em spray* whose patent number is PI0604112-4.

7b. The IPC of this patent are A46B 11/00 and A46B 13/04.

7c. A46B 11/00: Brushes with reservoir or other means for applying substances, for example, paints, pastes, water (driven brush bodies A46B 13/00; applying liquids or other fluent materials to surfaces by liquid carrying members in general, e.g., by pads, B05C 1/00, B05D 1/28).

    A46B 13/04: Brushes with driven brush bodies (power-driven toothbrushes A61C 17/16) with reservoir or other means for supplying substances.

8. To resolve this exercise, select *Assignee* as the search option of Derwent Basic and type *Loreal* **OR** *L'Oreal* in the search field. Below the search field, we select the period in which the patents were filed or granted. Below the main search field, where it reads *Timespan*, we select the option from 1963–1966 to 2017 and click *Search*. The number of results corresponds to the number of L'Oreal patents. This company holds 12,599 patents until the year 2017.

9. For the resolution of this exercise, the *Inventor* search option is selected in Derwent Basic. The link *Select from index* will appear where we click to access the next page that supports the technical search by inventor names. In the search field below where it reads *Click on the letter or type a few letters from the beginning of the name to browse alphabetically by inventor,* we type *Kubota* and click *Move to* button. Using the *Next* and *Previous* buttons, we found the names KUBOTA L and KUBOTA L T, which are selected by clicking the corresponding *Add* button. By clicking the *OK* button, we return to the main search page where we see the configuration *KUBOTA L* **OR** *KUBOTA L T*. Clicking *Search*, 21 results will appear. This number corresponds to the number of patents of researcher Lauro Tatsuo Kubota until the year 2017.

10. To resolve this exercise, select the patent number as search option of Derwent Basic and type WO2008034207 in the search field. The resulting patent is titled *Obtaining silver nanoparticle used in human and animal healthcare, e.g. in combating hospital infections, by using biosynthetic method where fungus extract has nitrate reductase activity and in the presence of anthraquinone derivatives.* The involved assignees are the State University of Campinas, Mogiana Organization of Education and Culture, and Mogi das Cruzes University.

11. To solve this exercise, we opened the Markush structure editor by choosing the *Exact search* option. The drawn molecule and the requested options are shown in Figure 4.4.

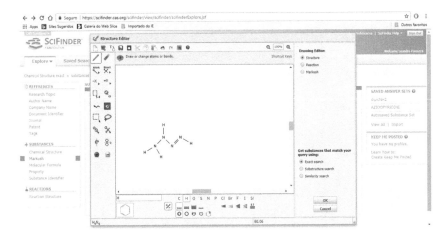

Figure 4.4 To draw chemical structures. (Source: Image from SciFinder used with permission from CAS, a division of the American Chemical Society.)

Clicking *OK* and *Search,* the substance options related to tetrazene will appear. Select all options, click *Get references,* select *All substances,* and click *Get.* The found results are refined by document type (patent) and publication year (1992). The title of the patent is *Detonator mixture for percussion caps.* The assignees of the patent are *Intreprinderea Mecanica Sadu, Bumbesti-Jiu,* and *Rom.*

12. For the resolution of this exercise, we open the editor of Markush structures choosing the *Exact search* option. The drawn molecule and the requested options are shown in Figure 4.5.

Clicking *OK* and *Search,* the substance options related to this pharmaceutical active compound will appear. Select all options, click *Get references,* select *All substances,* and click *Get.* The found results are refined by document type (patent) and publication year (1997). The pharmaceutical active compound is Sildenafil (Viagra), filed by the company Pfizer. The title of this patent is *Process for preparation of Sildenafil by cyclization.*

13. To solve this exercise, we opened the Markush structure editor by choosing the *Exact search* option. The drawn molecule and the requested options are shown in Figure 4.6.

Clicking *OK* and *Search,* the substance options related to this pharmaceutical active compound will appear. Select all options, click *Get references,* select *All substances,* and click *Get.* The found results are refined by document type (patent) and publication year (1923). The name of the compound is diazodinitrophenol, due to its important explosive properties. The patent number is US1460708.

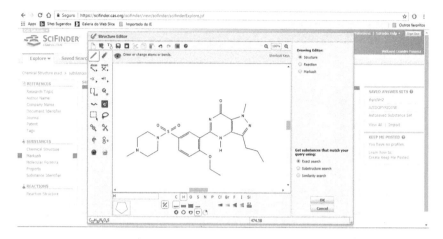

Figure 4.5 To draw chemical structures. (Source: Image from SciFinder used with permission from CAS, a division of the American Chemical Society.)

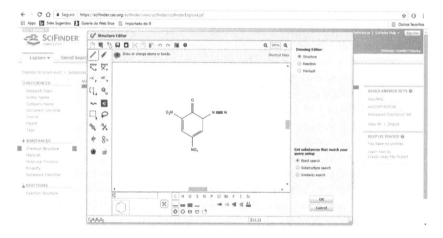

Figure 4.6 To draw chemical structures. (Source: Image from SciFinder used with permission from CAS, a division of the American Chemical Society.)

14a. Using Google Web, we typed the keywords *gary usa pop music august 29 1958*. The famous artist is Michael Jackson.

14b. For the resolution of this exercise, we entered the name of this artist in the field *Term 1* as *Jackson Michael J*, and we selected the search option *Title* in *Field 1*. When clicking *Search*, USPTO site generates more than one result. From this total, the only patent filed in 1993 by the artist Michael Jackson is US5255452 titled *Method and means for creating anti-gravity illusion*.

15. To resolve this exercise, enter the patent number in *Term 1* field as <u>*7366979*</u> and select the *Patent number* search option in *Field 1*. When you click *Search*, USPTO site generates a single result. According to the found patent, the celebrity is the director Steven Spielberg. The found patent is titled *Method and apparatus for annotating a document*.

16. For the resolution of this exercise, we enter the title of the patent in the field *Term 1* as <u>*Remote control device with gyroscopic stabilization and directional control*</u>, and we selected *Title* in *Field 1* as search option. When clicking *Search*, the site generates a single result. According to the found patent, the famous inventor is Jamie Hyneman, one of the presenters of Mythbusters series.

17. By clicking *Smart search* on the Espacenet net website, enter the keywords <u>*graphene oxide filter toxicity*</u>. From the obtained results, the document regarded to this concept is the patent titled *Graphene oxide filters and methods of use* whose patent number is US2013330833 or any other document included in its patent family. The involved inventors are Oscar N. Ruiz, K. Shiral Fernando, and Christopher Bunker.

18. Clicking *Smart search* on the Espacenet website, enter the number US2013074735. The found document is the patent titled *Process for the production of violacein and its derivative deoxyviolacein containing bioactive pigment from chromobacterium SP (MTCC5522)*. The involved inventor is Bhaskaran Krishnakumar.

19. For the resolution of this exercise, we type <u>*photovoltaic hybrid electrolysis cell*</u> in the field *All* (Common fields) and <u>*Berlin*</u> in the field *Inventor city* (*Inventor fields*) and select as search options documents filed in the last 20 years classified as EP documents and German Patents—available at the bottom of the page. The title of the patent is *Photovoltaic Hybrid Electrolysis Cell*, whose patent number is EP2817437. The inventors are Sebastian Fiechter, Rutger Schlatmann, and Bernd Stannowski.

20. To resolve this exercise, type <u>*Edward Van Halen*</u> in the *Inventor* field (Inventor fields) and select the US patents and US patent applications search options at the bottom of the page. Documents filed in all years are considered by selecting *All years*.

20a. Eddie holds four US patents, according to the results shown in the search carried out following the reasoning above.

20b. The oldest patent, filed in July 30, 1985, is titled *Musical Instrument support* and is filed with the patent number 4656917.

20c. The patent published in 2004 is titled *Stringed instrument with adjustable string tension control*, whose patent number is US20040159203.

21. To solve the exercise, we typed
<u>*nano gaps graphene*</u> in the *All* field (Common fields);
<u>*12/31/2014*</u> to <u>*01/01/2016*</u> in the *Filing Date* field (Date fields);
<u>*Oxford*</u> in the *Assignee* field.

For the search, were considered only US patents and US patent applications from all years. The disclosed patent is titled *Method for forming nano-gaps in graphene*, whose patent number is US20170145483. The involved inventors are George Andrew Davidson Briggs and Jan Andries Mol.

chapter five

Use of advanced patent search

Advanced patent searches are powerful tools for accurate investigation of patent documents in the context of a particular topic. In addition, given its importance, it is used by several patent offices around the world in patentability analysis processes and by researchers who plan the development of a novel product or method and use this kind of search to discover existing inventions and know how their idea is differentiated in relation to what exists in the state of the art. An advanced search performed according to a previous search strategy can generate, in most cases, results covering documents not found in a basic search using the same strategy. More specifically, the search for documents by the basic and advanced methods using the same strategy generates, respectively, "n" and "m" documents, where "m" > "n" and "n" is contained in "m" with few exceptions. This is due to the greater availability of search tools by the advanced method. In particular for advanced search databases, some of the key tools that deliver accurate and quality results are the joint operators and the wildcard characters. In general, each database covers its own set of search tools and will be specifically studied later. As with basic search, the advanced search method through Derwent and Orbit also comprises Boolean operators. Therefore, to better understand the presented databases below, we suggest the reader to remember the use of Boolean operators through a brief reading of the introduction of chapter 3 about basic searches.

5.1 Derwent advanced search

Derwent is accessed through the electronic page https://webofknowledge.com and by clicking the tab where it reads *Select a database* and selecting Derwent Innovations Index. We will automatically access the Derwent basic search page. From the latter page, we access the Derwent advanced search by clicking *Advanced search*. The main search page of Derwent advanced search is shown in Figure 5.1, where only one search field, designated herein as the main search field, is observed.

In advanced searches, the search options (as for basic searches) are designated as *Field tags* (displayed in the right corner of the screen as TS, TI, AU, etc.) and used to indicate in which part of the patent document we expect the keywords to be searched. In the context of Derwent advanced search, the search command follows a new logic. More specifically, the *Field*

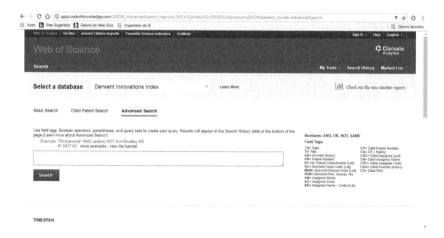

Figure 5.1 Main search field page of Derwent advanced search. (Source: Web of Science™ images are reproduced with permission from Clarivate Analytics © Clarivate Analytics 2018. Web of Science™, InCites™, Journal Citation Reports™, Essential Science Indicators™, Endnote™, Publons™, Clarivate Analytics™, and the logo of Clarivate Analytics are trademarks of Clarivate Analytics and its affiliated companies and are shown with permission. All rights reserved.)

tag is indicated by a code followed by an equal sign with keywords inserted in parentheses, as noted below. Considering a hypothetical *Field tag* of FT code, the keywords *keyword1* and *keyword2* will be inserted as follows:

$$FT = (\underline{keyword1} \textbf{ AND } \underline{keyword2})$$

Note that two combined keywords are in parentheses. In Derwent advanced search, the combination of at least two keywords must be enclosed in parentheses. Note the configuration below:

$$FT = \underline{keyword1} \textbf{ AND } \underline{keyword2}$$
$$\text{(Invalid command)}$$

The above configuration is invalid because the parentheses are absent and must be entered. Note, for example, that in the hypothesis of searching with only one hypothetical keyword *keyword1*, the use of parentheses is optional, so that the FT = *keyword1* and FT = (*keyword1*) configurations are equally valid and similar. In searches where two groups of keywords are used, parentheses can be entered in a logical way to be distinguished by the program. As an example, there are two groups of keywords below:

$$(\underline{keyword1} \textbf{ OR } \underline{keyword2})\text{---group 1}$$
$$(\underline{keyword3} \textbf{ OR } \underline{keyword4})\text{---group 2}$$

We could combine the above two groups using any Boolean operator within a certain *Field tag* of code FT, as follows:

$$FT = ((\underline{keyword1} \textbf{ OR } \underline{keyword2}) \textbf{ NOT } (\underline{keyword3} \text{ OR } \underline{keyword4}))$$

The both external parentheses (at the right and at the left) define in which *Field tag* the keywords *keyword1*, *keyword2*, *keyword3*, and *keyword4* will be searched for. The pair of parentheses surrounding the words *keyword1* and *keyword2* defines the group 1, and the pair of parentheses surrounding *keyword3* and *keword4* defines the group 2. Illustratively, the above search is configured to search for documents containing the words *keyword1* or *keyword2* and exclude patents containing the words *keyword3* or *keyword4*. Considering a practical example, we can consider a search in which keywords are contained in the title of the documents, whose *Field tag* is TI. Assuming we want to find patents that mention gold nanoparticles, silver nanoparticles, gold nanocomposites, or silver nanocomposites, one of the possibilities could be as follows:

$$TI = ((\underline{silver} \text{ OR } \underline{gold}) \textbf{ AND } (\underline{nanoparticle} \text{ OR } \underline{nanocomposite}))$$

The above search strategy possibly yields results in which the documents comprise some of the nanomaterials we plan to find.

Knowing the logic involved in Derwent advanced search, it is necessary to understand the possible *Field tags* of the program, which is explained as follows:

TS = Topic: It searches for documents in which the entered keywords are contained in the title or abstract of the patent;

TI = Title: It searches for documents in which the keywords entered are contained in the title of the patent;

AU = Inventor: It searches for documents in which the entered names are contained in the field of inventors' name of the patent. In this case, there is a list of inventors in the link *Index* next to where it reads *AU = Inventor* (right corner of the screen) where the name of the inventor (s) can be grouped. The logic of choice is similar to Derwent basic search, which can be found at the item III of section 3.3;

PN = Patent number: It searches for documents by patent number;

IP = Int. patent classification: It searches for documents by the international patent classification (IPC) whose list of IPC options in the link *List* is contained next to where it reads *IP = Int. patent classification* (right corner of the screen);

DC = Derwent class code: It searches for documents by the Derwent class code whose list of options in the link *List* is contained next to where it reads *DC = Derwent class code* (right corner of the screen). For a better understanding of this *Field tag*, read item VI of section 3.3;

MAN = Derwent manual code: It searches for documents in the Derwent manual code whose list of options in the link *List* is contained next to where it reads *MAN = Derwent manual code*. For a better understanding of this *Field tag*, read item VII of section 3.3;

PAN = Derwent prim. access. no.: It searches for documents by Derwent primary accession number (PAN). Each patent document registered in Derwent holds a specific PAN, allowing this type of search when there is no knowledge about the patent number of a patent of interest or in the absence of any other information that favors a precise and direct search;

AN = Assignee name: It searches documents by assignee name;

AC = Assignee code: It searches documents by assignee code. The assignee code is presented as a more precise way of searching for this option since a given assignee can present several name variants in several patent documents—it is similar to a synonym group for the same assignee. The program understands the inserted code as synonyms for a given assignee. For example, the company 3M holds patents in which its name can be observed as 3M Brazil LTDA, 3M France SA, and 3M Cogent, Inc., among others. The code referring to 3M has been designated by Derwent as *MINN-C*, that is, by entering AC = *MINN-C* in the search field, the final result will comprise documents whose 3M name will be displayed in at least two different manners (3M Brazil LTDA, 3M France SA, 3M Cogent, Inc., or other possibilities). To find out the code of a particular assignee, just click the link *List* next to where it reads *AE = Assignee Name + Code* (right corner of the screen). On the next page, enter the name of the assignee (3M, Basf, Unilever, Oxford, Harvard, etc.) in the search field below where it reads *Enter text to find a name or code containing or related to the text* and then click *Find*. The list with the assignee code (or codes) related to the respective assignee names will appear and can be copied in the search field of the previous page as AC = (*found code*);

AE = Assignee name + code: It searches documents by the assignee name in addition to the assignee code whose list of options in the link *List* is contained next to where it reads *AE = Assignee Name + Code* (right corner of the screen). To find out the code of a given assignee, read the *Field tag* above (AC = Assignee code);

CP = Cited patent number: It searches documents citing the inserted patent number in that *Field tag*. More specifically, when we enter, for example, CP = *US7033415*, the generated result covers patent documents citing the patent US7033415;

CX = CP + family: It searches for patent documents and their respective patent families that cite the inserted patent number in that *Field tag*, such as the *Field tag* CP above. When we enter, for example, CX = *US7033415*, the generated result encompasses patent documents (and their patent family) which cite the patent US7033415;

CAC = Cited assignee: It searches for patent documents that cite the entered assignee in that *Field tag*. The name can be entered as a code or as the own name. For example, the 3M company can be entered as CAC = _3M_ or CAC = _MINN-C_ since its code is MINN-C. The list of codes is in the link *List* next to where it reads *CAC* = *Cited assignee* (right corner of the screen). To find out the code, read about the *Field tag* AC = Assignee code, whose logic is similar;

CN = Cited assignee name: It searches for patent documents that cite the entered assignee in that *Field tag*. Only the assignee name is entered. For example, the 3M company can be inserted as CN = _3M_;

CPC = Cited assignee code: It searches for patent documents that cite the entered assignee in this *Field tag*. The name is entered only by the code. For example, the 3M company can be entered as CAC = _MINN-C_ since its code is MINN-C. The list of codes is in the link *List* next to where it reads *CAC* = *Cited assignee* (right corner of the screen). To find out the code, read about the *Field tag* AC = Assignee code, whose logic is similar;

CAU = Cited inventor: It searches for patent documents that cite the inserted inventor in that *Field tag*. In this case, there is a list of inventors in the link *Index* next to where it reads *CAU* = *Cited inventor* (right corner of the screen) where the name of the inventor (s) can be grouped. The logic of choice is similar to Derwent basic, which can be found in item III of section 3.3;

CD = PAN: It searches for documents that cite the inserted PAN in that *Field tag*. More specifically, when we type, for example, CD = _2004-795171_, the generated result covers patent documents that cite the patent whose PAN is 2004-795171 (the patent relating to this number is the same mentioned document in the examples of *Field tags* CP and CX, US7033415);

In the context of advanced searches, several searches can be managed and combined through a panel designated as *Search history*, where we can analyze each one of them as well as the number of found documents, the added keywords and the inserted *Field tags*. Each search included in *Search history* is also designated as *set*—a *Search history* with five searches contains five sets. The panel is located in the bottom corner of the screen. For your better understanding, we use the main search field to request the following searches, in the order below:

TS = (_silica_ _nanoparticle_)
TS = (_silica_ **AND** _nanoparticle_)
TS = _nanoparticle_
TS = (_nanoparticle_ **NOT** _particle_)

The above request generates the *Search history* containing the grouped sets as shown in Figure 5.2.

Search history contains columns showing each numbered set (set #1, set #2, etc.) in the column *Set*, its respective number of found documents in the

Figure 5.2 Panel of the *Search history* of Derwent advanced search. (Source: Web of Science™ images are reproduced with permission from Clarivate Analytics © Clarivate Analytics 2018. Web of Science™, InCites™, Journal Citation Reports™, Essential Science Indicators™, Endnote™, Publons™, Clarivate Analytics™, and the logo of Clarivate Analytics are trademarks of Clarivate Analytics and its affiliated companies and are shown with permission. All rights reserved.)

column *Results*, and the keywords displayed in the middle column. The *Combine sets* column allows you to combine any of the sets using the **AND** or **OR** operator. For this purpose, simply select the searches of interest, select **AND** or **OR** and click *Combine*, generating a fifth search designated as set #5. As an example, we have selected the searches #4, #2, and #1 and combined the three sets with the **OR** operator by selecting it in the *Combine sets* column and clicking *Combine*. Set #5 will display the results and indicate the combination as #4 **OR** #2 **OR** #1. Another way to combine searches in this way is to type #4 **OR** #2 **OR** #1 in the main search field—you do not need to enter *Field tags* for combination involving only the sets. The combination through the latter method can be advantageous against the column selection method when we want to insert the **NOT** operator since the latter is not available in the *Combine sets* column. As an example, we could exclude from set #3 all the found documents in set #2 by typing #3 **NOT** #2 in the main search field. Another possibility for *Search history* management is to combine one or more keywords with one or more sets through the main search field. Suppose we want to combine the set #3 with the words *gold* or *golden*. In the presence of keywords, it is necessary to indicate a *Field tag*. Therefore, we will have the possible search configuration:

$$TS = ((\underline{gold} \textbf{ OR } \underline{golden}) \textbf{ AND } \#3)$$

In the above configuration, we have the combination of two keywords with set #3—note that in the combination of sets, the *Field tags* are not required, while the keyword–set combination demands their use. Another advantage of combining searches by the main search field is the simultaneous use of more than one Boolean operator, which is not possible by the column-by-column method. Note that in the latter case, searches can only be combined with the **AND** operator in one situation or with the **OR** operator in another context.

It is notorious that the use of Boolean operators follows a line of reasoning similar to the basic searches and can be used by the two mentioned methods above. On the main page of Derwent advanced search, it is noted that another Boolean operator available in the Web of Science is the **SAME** operator. This operator is used to search for addresses on the Web of Science Core Collection platform (scientific articles) and has a function similar to the **AND** operator on the Derwent platform. The latter, in turn, is inserted implicitly between typed keywords in the absence of Boolean operators. Note the sets #1 and #2 in Figure 5.2. Note that the used keywords in both cases are similar, differing only in the fact that in the first and second searches, the **AND** operator is absent and present, respectively. Therefore, set #1 contains implicit **AND** operator between the words *silica* and *nanoparticle* so that the number of found documents is similar to set #2. *Search history* can be saved by clicking *Save history/Create alert* button and entering a newly created login and password. A saved search history can also be opened on another occasion by clicking *Open saved history* and entering a login and password. The search set can be partially or totally deleted via the *Delete* button in the *Delete set* column. For the total deletion of the sets, select *Select all* and click *Delete*. To partially exclude the sets, simply select them through the *Delete sets* column and click *Delete*.

Next, we will study two important tools that characterize one of the advanced search differentials and improve search accuracy. The first instrument is the joint operators, and the second is the wildcard characters.

5.1.1   Derwent advanced joint operators

Joint operators, common in advanced searches, are tools applied to combine at least two keywords that determine the maximum distance that one word should be from the other. The distance variation is established by the number of words that lie between them. More specifically, it is possible to insert a joint operator between the words *keyword1* and *keyword2* and establish the maximum number of words that separate the word *keyword1* from the word *keyword2* or vice versa. When we say that both are separated by even one word of distance, regardless of the typed order, we would have the following possibility:

(*keyword1* keyword-x *keyword2*) or (*keyword2* keyword-x *keyword1*)

or

(*keyword1* *keyword2*) or (*keyword2* *keyword1*)

Note that there is at most one *keyword-x* that separates *keyword1* and *keyword2*, not necessarily in the order in which it is typed, that is, *keyword1* can be to the left or to the right of the word *keyword-x* and *keyword2* can be to the left or to the right of the word *keyword-x*. There are joint operators that separate words that can be searched necessarily in the order they are typed (as in the case of Orbit, which we will study later). In the situation of typing *keyword1* OPERATOR *keyword2*, this joint operator requires that the words *keyword1* and *keyword2* are necessarily searched in the order in which they are typed and with even one word away. In this context, we would have the following possibilities:

keyword1 keyword-x *keyword2*

or

keyword1 *keyword2*

Note that the word *keyword1* is necessarily on the left and *keyword2* is necessarily on the right because they were typed in the order (left) *keyword1* OPERATOR *keyword2* (right). In the situation of applying this same joint operator (words searched in the typed order) with up to two words of distance, we would have the following possibility:

keyword1 keyword-x keyword-y *keyword2*
(two random keywords between *keyword1* and *keyword2*)

or

keyword1 keyword-x *keyword2*
(one random keyword between *keyword1* and *keyword2*)

or

keyword1 *keyword2*
(no keyword between *keyword1* and *keyword2*)

Note that the words *keyword-x* and *keyword-y* correspond to the maximum number of words that can separate *keyword1* and *keyword2*, that is, the possibility of applying joint operator with up to two words away (in direct or random search orders) also covers the possibility of only one or no word being found between *keyword1* and *keyword2*. In Derwent advanced search, the joint operator used is the **NEAR/x** operator, where **x** is the number of words of distance. In this way, **NEAR/3** and **NEAR/5**, for example, are examples of up to three and five words away between two keywords, respectively. The **NEAR/x** operator searches for the keywords in the order they are typed or in the reverse order. As an example, the TS = (*electric* **NEAR/3** *car*) search can find documents in which the words *electric* or *car* are typed on the left or right.

5.1.2   *Derwent advanced wildcard characters*

Wildcard characters are symbols inserted in the prefix, suffix, middle, or place of a character (or letter) in a given keyword, which indicate the possibility of none, one, or more than one character. The application of this tool allows us to find keywords with similar radicals (*metal, metallurgy, nonmetal,* etc.), words in the singular or plural forms, and possible spelling errors. The act of inserting a wildcard character into a word is known as word truncation. In Derwent, the wildcard characters are applied in the middle and/or in the word suffix and are shown below:

    a. Character *: It covers any group of characters (or letters), including none character. In the truncation of the word *nanoparticle* with the * character, we will have, among the numerous possibilities, the following examples:

*nanopartic** = *nanoparticle* **OR** *nanoparticles* **OR** *nanoparticulate* **OR** *nanoparticulates* **OR** *nanopartic* **OR** etc.

By analyzing the above example, it can be understood that typing *nanopartic** is similar to typing *nanoparticle* **OR** *nanoparticles* **OR** *nanoparticulate* **OR** *nanoparticulates* **OR** *nanopartic* **OR** any other possible word derived from the *nanopartic* radical not otherwise provided for therein. Let us analyze the number of involved characters in each of the possible words above:

*nanopartic** (1 wildcard character) = *nanoparticle* (2 characters) **OR** *nanoparticles* (3 characters) **OR** *nanoparticulate* (4 characters) **OR** *nanoparticulates* (5 characters) **OR** *nanopartic* (none character)

According to the definition and above example, we know, therefore, that the wildcard character * inserted in the word covers none or at least one character in the place where it is inserted. In an advanced search comprising the word *nanopartic** to be searched in the title and abstract of patents, the configuration would be TS = (*nanopartic**).

    b. Character ?: It necessarily covers one character (or letter). In the truncation of the word *polymerization* with the character ?, we will have, among the several possibilities, the following examples:

pol?merization = *polymerization* **OR** *polimerization* **OR** *polomerization* **OR** *polumerization* **OR** etc.

By analyzing the above example, it can be understood that typing *pol?merization* is similar to typing *polymerization* **OR** *polimerization* **OR** *polomerization* **OR** *polumerization* or any other possible word derived from radical *polymerization* not otherwise provided for therein. Let us check the number of involved characters in each of the possible words above:

pol?merization (1 wildcard character) = *polymerization* (1 character) **OR** *polimerization* (1 character) **OR** *polomerization* (1 character) **OR** *polumerization* (1 character)

As noted, in all of the above cases, the wildcard character necessarily comprises necessarily one character. In an advanced search
comprising the word *pol?merization* to be searched in the title and
abstract of patents, the configuration would be TS = (*pol?merization*).

c. Character $: It covers one or none character (or letter). In the trunca-
tion of the word *polymerizatio*n with the $ character, we will have,
among the several possibilities, the following examples:

polymeri$ation = *polymerization* **OR** *polymerisation* **OR** *polymeriation* **OR**
etc.

By analyzing the above example, it can be understood that typing
polymeri$ation is similar to writing *polymerization* **OR** *polymerisation*
OR *polymeriation* or any other possible word derived from the *polymerization* radical not otherwise provided for therein. Let us check the
number of involved characters in each of the possible words above:

polymeri$ation (1 wildcard character) = *polymerization* (1 character) **OR**
polymerisation (1 character) **OR** *polymeriation* (none character)

According to the definition and above example, we know, therefore, that the wildcard character $ inserted in the word covers one
or none characters in the place where it is inserted. In an advanced
search comprising the word *polymerisation* to be sought in the
title and abstract of the patents, the configuration would be TS =
(*polymeri$ation*).

We have studied the differences in the use of wildcard characters *, ?, and
$ of Derwent advanced search which, as noted, provide important results
in the context of spanning a greater number of keywords and their variants by typing only one in the main search field. Fortunately, for greater
accuracy and quality of this process, the three wildcard characters can be
simultaneously used in the same word. Considering the word polymerization truncated as pol?meri$at*, we would have several possibilities of
derived keywords, which shows the importance of truncation with regard
to the widest possible range of documents related to the given topic, reducing the risk of relevant documents not found in a search and increasing
the quality and accuracy of the work.

After studying all the search tools of Derwent advanced search, we
can understand the analysis of documents and the structuring of search
results that follows reasoning similar to Derwent basic. The found results
in each of the sets of a *Search history* can be accessed by clicking the link of
any of the numbers of found documents in the *Results* column. By clicking
any of them, we will go to the Derwent result page. Figure 5.3 shows the
result page of set #1 from the shown *Search history* in Figure 5.2.

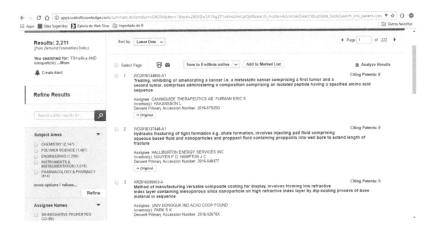

Figure 5.3 Search result page of Derwent advanced search. (Source: Web of Science™ images are reproduced with permission from Clarivate Analytics © Clarivate Analytics 2018. Web of Science™, InCites™, Journal Citation Reports™, Essential Science Indicators™, Endnote™, Publons™, Clarivate Analytics™, and the logo of Clarivate Analytics are trademarks of Clarivate Analytics and its affiliated companies and are shown with permission. All rights reserved.)

Exploring the above page, we have the number of found results displayed in the upper left corner of the screen as well as a brief summary of the keyword combination in *You searched for*. There are, in the context of Derwent advanced search, refinement options shown in tabs in the column in the left corner of the screen. Among the several possibilities, the number of found documents can be better refined by subject areas, assignee names, inventors, IPC codes, Derwent class codes, and Derwent manual codes. To refine the results for a given option, simply click that option and select the items of interest that automatically appear after clicking the tab. After the choices, just click *refine*. Note that for each refinement tab, there is the link *more options/values …* that displays additional refinement items. In succession, several refinement strategies using the left-hand column flaps can be performed until reasonable results are achieved for analysis.

The central screen presents the main results covering the patents, which can be accessed by clicking the title of the document for viewing the abstract via Derwent or in the button *original* for direct access to the patent content. The abstract view opens the page shown in Figure 5.4.

In basic searches, the process is performed in a single step in most databases and therefore does not require a robust search strategy. If there was a *Search history* in basic searches, the number of observed sets would be only one. In advanced searches, in turn, several searches designated as sets are strategically performed, combined, and viewed in the *Search history* panel. Therefore, the elaboration of a prior focused search strategy for an effective

Figure 5.4 Detailed information of a patent through Derwent advanced search. (Source: Web of Science™ images are reproduced with permission from Clarivate Analytics © Clarivate Analytics 2018. Web of Science™, InCites™, Journal Citation Reports™, Essential Science Indicators™, Endnote™, Publons™, Clarivate Analytics™, and the logo of Clarivate Analytics are trademarks of Clarivate Analytics and its affiliated companies and are shown with permission. All rights reserved.)

combination and execution of accurate searches is crucial. Understanding how Boolean operators synchronize with joint operators and wildcard characters is also strategic from the point of view of advanced searches. Aiming to consolidate and apply the concepts studied so far, a case study was elaborated whose objective is to search for patent documents related to a theme using the available tools and following a search strategy elaborated by our group. It is important to understand that the proposed logic below is one of the innumerable ways in which an analyst can devise a search strategy and act toward analysis. In this way, the intention is to show how tools can be synchronized and how search refinements can be strategically performed to solve problems of significant number of documents that need to be reduced to a group of relevant patents within the proposed theme.

5.1.3   Derwent search strategy

In this case study, we suggest to the reader to clear the *Search history* and follow the next steps in their computer for a better learning. The observed numbers in this case study come from a search conducted in 2017, so that slight numerical differences between the reader's search and those shown here may occur due to the continuous publication of patents and constant updating of the database. The search proposal is as follows:

Find some patent document about silica nanoparticles with polyaniline or polypyrrole for electronic applications.

In a broader sense, we did a search for everything that already exists about nanoparticles in general by typing TS = (*nanopart?c**). We truncated the letter "I" in the word *nanoparticle* with the wildcard character ? to cover documents with possible misspellings (nanoparticle written as nanoparticle, for example). The wildcard character * was entered in the suffix to comprise words in the singular and plural forms or similar terms. The number of results was as follows:

#1) TS = *nanopart?c**: 53,613 documents

We did another wide search on all existing polymers by typing: TS = *pol?m**. We truncated the letter "y" with the wildcard character ? to cover possible misspellings (polimer, etc.) and inserted the wildcard character * in the suffix to comprise words in the singular and plural forms or similar words. The result was as follows:

#2) TS = *pol?m**: >100,000 documents
#1) TS = *nanopart?c**: 53,613 documents

From the above searches, we combined both of them with the **AND** operator to find out documents about polymeric nanoparticles in general. The result was as follows:

#3) #2 **AND** #1: 18,665 documents
#2) TS = *pol?m**: >100,000 documents
#1) TS = *nanopart?c**: 53,613 documents

Taking into account the possibility of finding patents whose words *pol?m** and *nanopart?c** are distant in the documents due to the presence of the **AND** operator, which may include patents of different themes, we refined search #3 using the **NEAR/5** operator so as to approximate *pol?m** and *nanopart?c** in up to five words away. In this way, the probability of comprehensiveness of all documents related to polymeric nanoparticles is increased. In this way, we entered TS = (*pol?m** **NEAR/5** *nanopart?c**) obtaining the following result:

#4) TS = (*pol?m** **NEAR/5** *nanopart?c**): 11,009 documents
#3) #2 **AND** #1: 18,665 documents
#2) TS = *pol?m**: >100,000 documents
#1) TS = *nanopart?c**: 53,613 documents

Note the refinement efficiency by applying a joint operator by comparing sets #3 and #4, whose reduction was from 18,665 to 11,009 (still large) documents. Another way to use the **NEAR/5** joint operator in searches #1 and #2 is by typing TS = (#1 **NEAR/5** #2), obtaining similar results as noted below:

#5) TS = (#1 **NEAR/5** #2): 11,009 documents
#4) TS = (*pol?m** **NEAR/5** *nanopart?c**): 11,009 documents
#3) #2 **AND** #1: 18,665 documents

#2) TS = *pol?m**: >100,000 documents
#1) TS = *nanopart?c**: 53,613 documents

The search #5 is still broad, so we could refine it by testing the distance between the words *pol?m** and *nanopart?c** for up to three words away using the **NEAR/3** operator. The result was as follows:

#6) TS = (# 1 **NEAR/3** # 2): 10,826 documents
#5) TS = (#1 **NEAR/5** #2): 11,009 documents
#4) TS = (*pol?m** **NEAR/5** *nanopart?c**): 11,009 documents
#3) #2 **AND** #1: 18,665 documents
#2) TS = *pol?m**: >100,000 documents
#1) TS = *nanopart?c**: 53,613 documents

The refinement was from 11,009 to 10,826 documents, still a wide value. Search #6 comprises possible documents related to polymeric nanoparticles of diverse chemical nature. In this way, we refined the results to search for documents about polymeric nanoparticles whose polymers are polyaniline or polypyrrole by typing TS = (*pol?an?lin** **OR** *pol?p?r$ol**). The truncation with wildcard character? in the words *polyaniline* and *polypyrrole* was performed to encompass possible misspellings in the letters "y" and "I" (*polyanyline*, *polianyline*, polianiline, polipyrrole, polypirrole, polipirrole, etc.) while the wildcard character $ was used in the word *polypyrrole* to encompass *polypyrrole* words written with only one "r" (*polypyrole*) or other possible errors. The use of the wildcard character* in both terms comprises words written in the singular and plural forms and other similar forms. The result was as follows:

#7) TS = (*pol?an?lin** **OR** *pol?p?r$ol**): 14,504 documents
#6) TS = (# 1 **NEAR/3** # 2): 10,826 documents
#5) TS = (#1 **NEAR/5** #2): 11,009 documents
#4) TS = (*pol?m** **NEAR/5** *nanopart?c**): 11,009 documents
#3) #2 **AND** #1: 18,665 documents
#2) TS = *pol?m**: >100,000 documents
#1) TS = *nanopart?c**: 53,613 documents

We then combined searches #7 and #6 with the **AND** operator to broadly analyze results for documents about polyaniline- or polypyrrole-based polymeric nanoparticles. The result was as follows:

#8) #7 **AND** #6: 440 documents
#7) TS = (*pol?an?lin** **OR** *pol?p?r$ol**): 14,504 documents
#6) TS = (# 1 **NEAR/3** # 2): 10,826 documents
#5) TS = (#1 **NEAR/5** #2): 11,009 documents
#4) TS = (*pol?m** **NEAR/5** *nanopart?c**): 11,009 documents
#3) #2 **AND** #1: 18,665 documents

#2) TS = *pol?m**: >100,000 documents
#1) TS = *nanopart?c**: 53,613 documents

A significant refinement of results from 14,504 to 440 documents was observed. However, we know that the **AND** operator randomly combines the keywords so that their distance in the document can be large. The distance of keywords, as we know, can lead to find even documents about different subjects. Thus, to improve the accuracy of set #8, we used the **NEAR/3** operator. We typed TS = (#7 **NEAR/3** #6) and obtained the following result:

#9) TS = (#7 **NEAR/3** #6): 440 documents
#8) #7 **AND** #6: 440 documents
#7) TS = (*pol?an?lin** **OR** *pol?p?r$ol**): 14,504 documents
#6) TS = (# 1 **NEAR/3** # 2): 10,826 documents
#5) TS = (#1 **NEAR/5** #2): 11,009 documents
#4) TS = (*pol?m** **NEAR/5** *nanopart?c**): 11,009 documents
#3) #2 **AND** #1: 18,665 documents
#2) TS = *pol?m**: >100,000 documents
#1) TS = *nanopart?c**: 53,613 documents

There was no change in the result. We refined the latter using the **NEAR/2** operator by typing TS = (#7 **NEAR/2** #6). From this search request, we have the following:

#10) TS = (#7 **NEAR/2** #6): 369 documents
#9) TS = (#7 **NEAR/3** #6): 440 documents
#8) #7 **AND** #6: 440 documents
#7) TS = (*pol?an?lin** **OR** *pol?p?r$ol**): 14,504 documents
#6) TS = (# 1 **NEAR/3** # 2): 10,826 documents
#5) TS = (#1 **NEAR/5** #2): 11,009 documents
#4) TS = (*pol?m** **NEAR/5** *nanopart?c**): 11,009 documents
#3) #2 **AND** #1: 18,665 documents
#2) TS = *pol?m**: >100,000 documents
#1) TS = *nanopart?c**: 53,613 documents

There was a slight refinement for 369 results concerning, possibly, accurate documents about polyaniline- and polypyrrole-based polymeric nanoparticles. We decided to refine the documents, still in significant quantity, through the application of the invention. Considering the electronic application, we thought about possible terms and synonyms that allude to electronic devices such as LED, network, electronic, computer, and so on. We then opened a new set by typing TS = (*LED* **OR** *network** **OR** *ele$tronic** **OR** comput*). The wildcard character * was entered in all words following the reasoning similar to the previous sets, and the wildcard character $ was entered in the word *electronic* for spanning misspelled words as in *electronic* written without the letter "c" (*eletronic*). The result of the search was as follows:

#11) TS = (*LED* **OR** *network** **OR** *ele\$tronic** **OR** comput*): >100.00
documents
#10) TS = (#7 **NEAR/2** #6): 369 documents
#9) TS = (#7 **NEAR/3** #6): 440 documents
#8) #7 **AND** #6: 440 documents
#7) TS = (*pol?an?lin** **OR** *pol?p?r\$ol**): 14,504 documents
#6) TS = (# 1 **NEAR/3** # 2): 10,826 documents
#5) TS = (#1 **NEAR/5** #2): 11,009 documents
#4) TS = (*pol?m** **NEAR/5** *nanopart?c**): 11,009 documents
#3) #2 **AND** #1: 18,665 documents
#2) TS = *pol?m**: >100,000 documents
#1) TS = *nanopart?c**: 53,613 documents

We have broadly combined searches #11 and #10 using the **AND** operator
to obtain the result below:

#12) #11 **AND** #10: 109 documents
#11) TS = (*LED* **OR** *network** **OR** *ele\$tronic** **OR** comput*): >100.00
documents
#10) TS = (#7 **NEAR/2** #6): 369 documents
#9) TS = (#7 **NEAR/3** #6): 440 documents
#8) #7 **AND** #6: 440 documents
#7) TS = (*pol?an?lin** **OR** *pol?p?r\$ol**): 14,504 documents
#6) TS = (# 1 **NEAR/3** # 2): 10,826 documents
#5) TS = (#1 **NEAR/5** #2): 11,009 documents
#4) TS = (*pol?m** **NEAR/5** *nanopart?c**): 11,009 documents
#3) #2 **AND** #1: 18,665 documents
#2) TS = *pol?m**: >100,000 documents
#1) TS = *nanopart?c**: 53,613 documents

In this situation, with 109 found documents in search #12, we are almost
reaching the group of documents about the proposed theme. We need to
improve the accuracy of this result by inserting the **NEAR/3** operator by
typing TS = (#11 **NEAR/3** #10). The result was as follows:

#13) TS = (#11 **NEAR/3** #10): 109 documents
#12) #11 **AND** #10: 109 documents
#11) TS = (*LED* **OR** *network** **OR** *ele\$tronic** **OR** comput*): >100.00
documents
#10) TS = (#7 **NEAR/2** #6): 369 documents
#9) TS = (#7 **NEAR/3** #6): 440 documents
#8) #7 **AND** #6: 440 documents
#7) TS = (*pol?an?lin** **OR** *pol?p?r\$ol**): 14,504 documents
#6) TS = (# 1 **NEAR/3** # 2): 10,826 documents
#5) TS = (#1 **NEAR/5** #2): 11,009 documents

#4) TS = (*pol?m* NEAR/5 *nanopart?c**): 11,009 documents
#3) #2 AND #1: 18,665 documents
#2) TS = *pol?m**: >100,000 documents
#1) TS = *nanopart?c**: 53,613 documents

There was no change in numbers. Set #13 may accurately cover documents related to polyaniline- or polypyrrole-based polymeric nanoparticles for electronic applications. In this case, the nanoparticles covered may be of any chemical nature, and as planned, we intended to find nanoparticles based on silica. In this way, we opened a search by typing TS = *silica* (set #14) and combined it with set #13 by applying the AND operator. The result was as follows:

#15) #14 AND #13: 14 documents
#14) TS = *silica*: >100.00 documents
#13) TS = (#11 NEAR/3 #10): 109 documents
#12) #11 AND #10: 109 documents
#11) TS = (*LED* OR *network** OR *ele$tronic** OR comput*): >100.00 documents
#10) TS = (#7 NEAR/2 #6): 369 documents
#9) TS = (#7 NEAR/3 #6): 440 documents
#8) #7 AND #6: 440 documents
#7) TS = (*pol?an?lin** OR *pol?p?r$ol**): 14,504 documents
#6) TS = (# 1 NEAR/3 # 2): 10,826 documents
#5) TS = (#1 NEAR/5 #2): 11,009 documents
#4) TS = (*pol?m* NEAR/5 *nanopart?c**): 11,009 documents
#3) #2 AND #1: 18,665 documents
#2) TS = *pol?m**: >100,000 documents
#1) TS = *nanopart?c**: 53,613 documents

We were able to find 14 documents possibly related to silica nanoparticles with polyaniline or polypyrrole for electronic applications. The *Search history* can be analyzed in Figure 5.5.

By clicking the number of found documents in set #14, we will access the Derwent search result electronic page and will find the patent of title *Composition, useful in an electronic device eg photoresistors, electrically conductive polymer doped with fluorinated acid polymer, non-conductive oxide nanoparticles, high boiling solvent and lower boiling solvent*. This document comprises the proposed invention in this case study.

As we have studied, advanced search presents itself as a powerful tool for analyzing the feasibility of new product or process development, patentability analysis, and robust bibliographic studies on a given topic. The domain of advanced search skills inserts research and development professionals and scientists in a differential context as it provides relevant and important information often not found in the basic search.

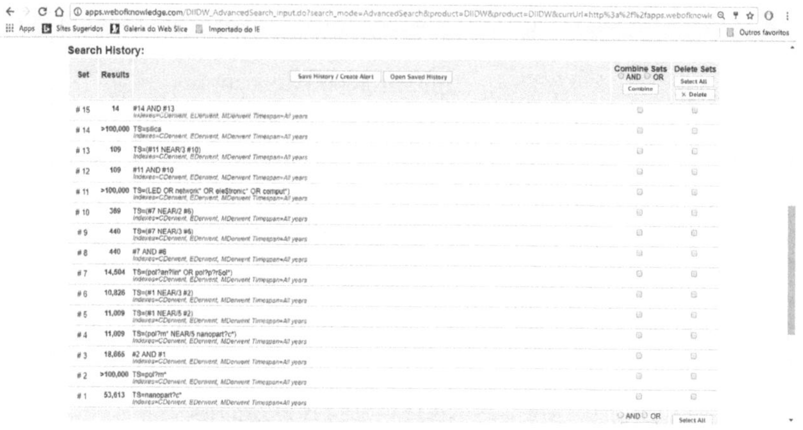

Figure 5.5 Panel of the *Search history* of the case study through Derwent advanced search. (Source: Web of Science™ images are reproduced with permission from Clarivate Analytics © Clarivate Analytics 2018. Web of Science™, InCites™, Journal Citation Reports™, Essential Science Indicators™, Endnote™, Publons™, Clarivate Analytics™, and the logo of Clarivate Analytics are trademarks of Clarivate Analytics and its affiliated companies and are shown with permission. All rights reserved.)

5.2　Orbit

The Orbit platform is a database that comprises the advanced search of patents as well as presents powerful tools of market analysis for creation of a business model. The website is accessed at www.orbit.com, and its home page requests a previously created login and password. After inserting this information, an easy search window initially appears. In this screen, the access of the advanced search is performed by clicking the >> button beside where it reads *Easy search* (upper left corner of the screen), which opens a column where the *Advanced search* link is observed. By clicking this latter link, we access the main electronic page of advanced search of Orbit, as observed in Figure 5.6.

The main advanced search page comprises the search fields *Keywords, Classifications, Names, Numbers (publication, application, and priority numbers), Dates, and Countries*. Although the search fields *Legal status* and *More fields* below the page are available, we will focus on those first ones because they cover the most advanced search strategies and are significantly specific (isolated problems).

5.2.1　Keywords topic

This search field involves the use and combination of keywords as we know it. In addition to being one of the most important of Orbit platform,

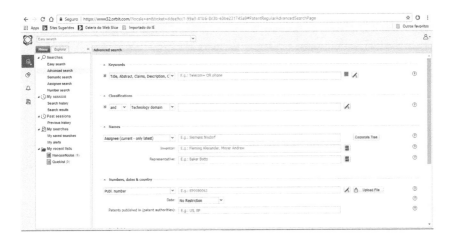

Figure 5.6 Main electronic page of advanced search of Orbit. (Source: © Questel "Orbit IPBI".)

the *Keywords* search field presents the Boolean operators **AND**, **OR**, and **NOT**; wildcard characters with different symbols, but with similar functions of those ones we studied in Derwent advanced search; and, interestingly, an infinity of several important and strategic joint operators. More specifically, while Derwent advanced search presents only one joint operator (**NEAR/x**), we will see the availability of six of these operators by Orbit, characterizing one of its search differentials.

Keywords are entered and combined in the blank space (or field) where *E.g.: Telecom+ OR phone* is read. In the case of typing two or more keywords in the absence of Boolean operators, the program understands the words as one side by side, necessarily in the typed order and up to one word away (read about this concept in the topic about joint operators furthermore). For example, by typing *electric car*, the program searches for both words in the order they are typed and in the presence or absence of only one random word between *car* and *electric*. Unlike Derwent advanced search, Orbit does not work with *Field tag* codes, and the addition of parentheses in the extreme left or extreme right is optional. However, organizing terms using parentheses is essential for recognizing the combination of keywords when more than one Boolean operator is used, according to the logic we already know. Keyword groups can also be combined by adding more than one search field by clicking the "+" sign next to where it reads *Title, Abstract, Claims* in the upper center corner of the screen. In this situation, we have two search fields (one above and one newly added below) where we can freely enter the keywords. The added search fields are automatically combined with an implicit **AND** operator. Orbit lets you work with up to six keyword search fields combined with implicit **AND** operators, as shown in Figure 5.7.

Figure 5.7 The possible added search fields in the topic *Keywords*. (Source: ©
Questel "Orbit IPBI".)

Search fields for additional keywords can be deleted by clicking the "–"
sign. Although there are no codes to indicate *Field tags* in the search fields,
these are present in the tab where *Title, Abstract, Claims* is read and characterize a differential feature of this program since while Derwent provides
options of search of keywords in the title and abstract, Orbit fully extends
the scope by searching them in Title, Abstract, Claims, Description, Object of
invention, Description, Object of invention, Advantages over prior art drawbacks, Independent claims, Concepts, and Full text. The analyst can select
all, some, or just one *Orbit Field tag*, whose description is briefly noted below:

Title: It searches keywords in the title of the patent;

Abstract: It searches for keywords in the patent abstract;

Claims: It searches for keywords in all patent claims;

Description: It searches keywords in the topic of detailed description of
  the invention of the patent document;

Object of invention: It searches for keywords in the topic of brief
  description the invention of the patent document;

Advantages over prior art drawbacks: It searches keywords in the paragraph of the patent where the differential and the advantages of the
  invention are mentioned in relation to the state of the art;

Independent claims: It searches for keywords only in independent
  claims of the patent;

Concepts: It searches for keywords within a group of words or short
  phrases extracted by Orbit from the patent document. The said
  group comprises key concepts summarizing the patent as a whole;

Full text: It searches keywords in the full patent document, especially in the title, abstract, description, and claims.

As mentioned, the Boolean operators **AND, OR**, and **NOT** can be used in Orbit. Therefore, it is necessary to understand the joint operators of this program, characterized here as another differential.

5.2.1.1 Orbit joint operators

Joint operators are tools applied to combine at least two keywords that determine the maximum distance that one word should be from another. The distance variation is established by the number of words that lie between them and can be arbitrarily chosen. The most precise definition of this kind of operator can be found in section 5.1.1. The joint operators of this advanced search program are shown below.

Operator _: Underline operator is used to search keywords together or separated by hyphen, parentheses, or space. By inserting, for example, this operator in the word polyalkyl as poly_alkyl, we would have the following search possibilities:

Poly_alkyl = polyalkyl OR poly alkyl OR poly (alkyl) OR etc.

Analyzing the above example, we have the words poly_akyl joined as polyalkyl and separated as poly alkyl and poly (alkyl).

Operator **nD**: **nD** joint operator is used between keywords to indicate the spacing of up to "*n*" keywords. The search does not take into account the order in which the terms are typed. See the examples below:

mesoporous **3D** *silica*

The above example searches for the words *mesoporous* and *silica* in up to three words away, not necessarily in the order in which they are typed, that is, both words may be to the right or left of the **3D** operator (randomly). Let us analyze another example:

silver **D** *nanoparticle* = *silver* **1D** *nanoparticle*

When $n = 1$ in **nD**, the program understands the **1D** operator as similar to the operator **D** (implicit $n = 1$), that is, **1D** = **D**. In the above example, the words *silver* and *nanoparticle* will be searched in up to one word away, not necessarily in the order in which they are typed. It is interesting to note that the **nD** operator in Orbit holds the same function as the **NEAR/x** operator in Derwent advanced search.

Operator **nW**: **nW** joint operator is used between the keywords to indicate the spacing of up to "*n*" keywords. The search takes into consideration the order in which the terms are typed. See the examples below:

graphene **5W** *oxide*

The above example searches for the words *graphene* and *oxide* distant in up to five words away, necessarily in the order in which they are typed, that is, the word *graphene* necessarily to the left of the **5W** operator and *oxide* mandatorily to the right. Let us analyze another example:

lipid **W** *membrane* = *lipid* **1W** *membrane* = *lipid* *membrane*

When $n = 1$ in **nW**, the program understands the operator **1W** as similar to the operator **W** (implicit $n = 1$), that is, **1W** = **W**. In the above example, the words *lipid* and *membrane* will be searched in up to one word away, necessarily in the order in which they are typed. Note that, however, by typing *lipid* *membrane* program implicitly and automatically inserts the **1W** operator between them.

Operator **F**: Joint operator **F** (Field) is used between the keywords to simultaneously find them in a certain field (Title, Abstract, Claims, Description, Object of invention, Description, Object of invention, or Advantages over prior art drawbacks). Let us analyze the example below:

Amorphous **F** *structure*

In the above situation, the program searches for both words simultaneously present in at least one of the fields mentioned above, regardless of the number of words existing between *amorphous* and *structure*.

Operator **P**: Joint operator **P** (Paragraph) is used between keywords to simultaneously find them in a certain paragraph. Let us analyze the example below:

crystalline **P** *solid*

In the above situation, the program searches for both words simultaneously present in at least one of the paragraphs of the patent, regardless of the number of words existing between *crystalline* and *solid*.

Operator **S**: Joint operator **S** (Sentence) is used between the keywords to simultaneously find them in a certain sentence. Let us analyze the example below:

glass **S** *network*

In the above situation, the program searches for both words simultaneously present in at least one of the sentences of the patent, regardless of the number of words between *glass* and *network*.

In general, operators **F**, **P**, and **S** can gradually refine a certain result since it decreases, respectively, the number of words between two terms.

5.2.1.2   Orbit wildcard characters

Orbit displays wildcard characters similar in function compared to those studied in Derwent advanced search. The main differences, as we will see below, are the symbol used and the fact that in Orbit it is possible to insert them in the prefix, middle (replacing or not a letter), or suffix of the word.

a. Character * or +: It covers any group of characters (or letters), including none character. In the truncation of the word *structure* with the wildcard characters * or+we will have, among the numerous possibilities, the following examples:

struct+ = *structure* **OR** *structuring* **OR** *structures* **OR** *structs* **OR** *struct* **OR** etc.

*struct** = *structure* **OR** *structuring* **OR** *structures* **OR** *structs* **OR** *struct* **OR** etc.

By analyzing the above example, it can be understood that typing *struct+* or *struct** is similar to typing *structure* **OR** *structuring* **OR** *structures* **OR** *structs* **OR** *struct* or any other possible word derived from the *struct* radical not provided beyond these. Let us analyze the number of involved characters in each of the possible words above:

struct+ (1 wildcard character) = *structure* (3 characters) **OR** *structuring* (5 characters) **OR** *structures* (4 characters) **OR** *structs* (1 character) **OR** *struct* (none characters) **OR** etc.

According to the definition and above example, we know, therefore, that the wildcard character+(or *) inserted in the word covers none or at least one character in the place where it is inserted.

b. Character #: It necessarily covers one character (or letter). In the truncation of the word *polypyrrole* with the wildcard character #, we will have, among the numerous possibilities, the following examples:

pol#pyrrole = *polypyrrole* **OR** *polipyrrole* **OR** *polopyrrole* **OR** *polupyrrole* **OR** etc.

By analyzing the above example, it can be understood that typing *pol#pyrrol* is similar to typing *polypyrrole* **OR** *polipyrrole* **OR** *polopyrrole* **OR** *polupyrrole* or any other possible word derived from *polypyrrole* radical not otherwise provided. Let us analyze the number of characters involved in each of the possible words above:

pol#pyrrole (1 wildcard character) = *polypyrrole* (1 character) **OR** *polipyrrole* (1 character) **OR** *polopyrrole* (1 character) **OR** *polupyrrole* (1 character) **OR** etc.

As noted, in all cases, the wildcard character # necessarily comprises one character.

c. Character ?: It covers one or none character (or letter). In the truncation of the word *cotton* with the wildcard character ?, we will have, among the numerous possibilities, the following examples:

cot?on = *cotton* **OR** *cotron* **OR** *coton* **OR** etc.

By analyzing the above example, it can be understood that typing *cot?on* is similar to typing *cotton* **OR** *cotron* **OR** *coton* or any other possible word derived from the *cotton* radical not otherwise provided for therein. Let us analyze the number of characters involved in each of the possible words above:

cot?on (1 wildcard character) = *cotton* (1 character) **OR** *cotron* (1 character) **OR** *coton* (none character) **OR** etc.

According to the definition and above example, we know, therefore, that the wildcard character ? inserted in the word *cot?on* covers one or none characters in the place where it is inserted.

We studied the differences in the use of wildcard characters */+, #, and ? from Orbit, which can also be simultaneously inserted into a certain word, as we noted in Derwent advanced search. It is important to point out that the possibility of prefix truncation characterizes a special advantage of this program, especially in the search for complex names of organic molecules. This technique will be better explained in the topic below.

5.2.1.3   Using operators and wildcard characters from Orbit

In this section, we will see how the use of the wildcard characters, especially as they are applicable in the word prefix, together with the application of joint operators, facilitates the patent search process, especially for searches involving names of organic molecules. For a broader context, we will exemplify this example with a search of documents referring to the molecule derived from Camptothecin, a chemotherapeutic agent. The IUPAC name of this molecule is (S)-4-ethyl-4-hydroxy-1H-pyrano[3',4':6,7] indolizino[1,2-b]quinoline-3,14-(4H,12H)-dione. Let us check one of the many truncation possibilities:

(S)-4-ethyl-4-hydroxy-1H-pyrano[3',4':6,7]indolizino[1,2-b] quinoline-3,14-(4H,12H)-dione

Terms such as [3',4':6,7], -(4H,12H)- and (S)-4- are specific and important in the designation of the molecule. However, Derwent and Orbit programs do not recognize them, and therefore, by typing the full specific name in the search field, error messages will appear. Focusing on solving this problem, let us analyze the terms in red below:

(S)-4-ethyl-4-hydroxy-1H-pyrano[3',4':6,7]indolizino[1,2-b]
quinoline-3,14-(4H)-dione

We will begin with the exclusion of the red terms and inclusion of joint operators as noted below:

Ethyl **5W** hydroxy **5W** pyrano **5W** indolizino **5W** quinoline **5W**
dione

We have applied **nW** operators to request the program a search for molecules whose words above are kept in the typed order. The spacing of $n = 5$ was inserted as a guarantee of the coverage of the hidden terms and their possible variations. We will finally apply the wildcard characters as follows:

+eth#l 5W h#drox+ 5W p#ran+ 5W indolizin+ 5W quinolin+ 5W
dion+

In general, wildcard characters # have been applied to cover possible misspellings of the letters "y," while the wildcards characters+in the suffix include errors and words in the singular/plural forms. Unlike what would be possible in Derwent advanced search, we have inserted the wildcard character+in the prefix of the molecule name (next to the word +eth#1) since the term (S)-4- can be covered.

So far, we have learned how to manage the Keywords topic and strategically apply the search tools. In later subjects, we will study the other topics present on this platform.

5.2.2   *Classifications topic*

This field allows searches using the following types of patent classifications: IPC, Cooperative Patent Classification, European Classification System, In-Computer-Only code and the Japanese patent classifications such as the File Forming Term (F) and the File Index Term (FI). These patent classifications can be chosen on the *Technology domain* tab. Besides this, there is the field where classifications can be included and combined using operators and wildcard characters. By choosing one of the search possibilities in *Technology domain*, the search field displays an example of including the patent classification. When choosing, for example, the IPC, the example *E.g.: C12N-001/21* will appear in the search field. It is notorious that patent classifications are significantly specific, and therefore, to make them broader, we could truncate them with wildcard characters. In the example of the C12N-001/21 classification, the truncation with the wildcard character+increases the search amplitude in the order C12N-001+ (any document comprising the C12N-001 classification), C12N+ (any document comprising the C12N classification), and C+ (any document comprising the C classification). The said versatility in the search of documents by the classification of patents confers a robust search profile (countless possibilities can be thought of). The above example is applicable for any classification of the *Technology domain* tab. It is important to remember that patent classifications can be combined using Boolean operators, as in the example (C12N-001+ **OR** A01N+ **OR** F41G-011/00 OR H+). As in the *Keywords* topic, the *Classifications* topic allows the analyst to add other search fields by patent classifications and combine them with **AND** or **OR** operators in the tabs where you initially read *and*. The maximum number of search fields in *Classification* is five. It is also possible to exclude search fields by clicking the "–" sign, just like in *Keywords*.

5.2.3   Names topic

This section allows the search of patents by assignee, inventor, and representative agent of patents (*representative*). It is important to note that the patent search in *Names* also allows the use of operators and wildcard characters.

5.2.4   Numbers, dates, and country search field

This section searches for patent documents in the following options and suboptions:

a. Numbers: The search field next to where it reads *Publ. number* searches documents by publication number and application number. These suboptions can be selected on the tab where *Publ. number* is read.
b. Date: The search field next to where it reads *date* searches documents by periods or dates including application date, priority date, and publication date, which can be selected in the tab where *No restriction* is read.
c. Patents published in (patent authorities): It searches for documents by patent authorities (US, EP, JP, etc.).

Orbit automatically combines all topics *Keywords, Classifications, Names, Numbers, Legal status* and *More fields* with implicit **AND** operator.

When searching for any of the presented options, the search request is performed by clicking *Search* in the lower corner of the screen or by pressing *Enter*, opening the next page referring to the results obtained, also called Search results.

The search result page from Orbit displays the title of each found patent, as well as the publication number, first application date, and assignee. The detailed information of each document can be accessed by clicking the link (title) of the patent of interest. The right corner of the screen consists of a quick information window that, in the first instance, shows the main figure of the patent. Other information for quick access can be arbitrarily chosen by clicking the "+" sign in the upper right corner of the screen. Among the available quick access information, the Claims, Description, and Citations, among other options, can be cited. The previous search page can be accessed again by clicking *Menu* in the upper left corner of the screen and clicking *Advanced search*. Below these options, you can access the Orbit *Search history*, which we will visualize the structure below, and the link *My saved searches* that redirects to the saved search page. The *Search results* page could still be saved by clicking *Save strategy* in the upper middle corner of the screen that opens a new window. The latter shows the search summary and requests a title. Next, click *Save* in the bottom corner, and finally, the search can be viewed in *My saved searches*. Accessing the page of saved searches, a history will appear. Each saved item can be executed, edited, or deleted by clicking *Execute*, *Edit* and *Delete*, respectively. By clicking *Search results* in the left column, we will return to the previous page to understand the detailed view of a certain patent by clicking its respective title. The detailed patent description page is shown in Figure 5.8.

The detailed description page of a patent in Orbit initially shows the bibliographic information of the document. The information is grouped in tabs available in the upper corner of the screen and comprises the tabs *Biblio, Claims, Description, Key content, Concepts, Fulltext, Kwic (keywords in context), Legal status, citations, and drawings*. Each group can be analyzed separately by clicking the respective tab. Orbit succinctly shows in the center of the screen the patent family containing all the covered documents (including the patent being analyzed) that can be fully accessed by

Figure 5.8 Detailed information page of a patent through Orbit. (Source: © Questel "Orbit IPBI".)

downloading its PDF version from the icon in the *Document type* column (first icon from left to right). It is important to note that in Orbit, the documents are available or translated automatically into English. By clicking *Translate* under the *Biblio* tab, the program provides other languages in which the documents can be translated.

The next study stage is to understand the management of your *Search history* by clicking *Menu* in the left corner of the screen and clicking *Search history*. The summary of the sets performed is displayed on the next electronic page, as shown in Figure 5.9.

As noted above, each Orbit Search history set can be accessed, edited, saved, and deleted by clicking, respectively, the links *Show results, Modify, Save,* and *Delete* on the right-hand corner of each line for a certain search. An alert for each search can also be created by clicking the link *Alert* in each set, where its frequency (monthly or weekly) is chosen. Each set of *Search history* can be combined through the search field where it reads *Combine strategies*. The combination of the searches is direct, that is, the number of sets to be combined and the operators (Boolean or joint) applied between them are freely typed. For example

$$(1 \textbf{ OR } 4 \textbf{ OR } 3) \textbf{ AND } (2 \textbf{ OR } 5)$$
$$(1 \textbf{ OR } 4 \textbf{ OR } 3) \textbf{ F } (2 \textbf{ OR } 5)$$
$$(4 \textbf{ AND } 5) \textbf{ NOT } 1$$

As for each set individually, the whole *Search history* can be saved by clicking *Save entire strategy* in the upper left corner. A new window asking for the title of the *Search history* will appear. Just click *Save* and access the content in *My saved searches* in the future.

Figure 5.9 Panel of the *Search history* of Orbit. (Source: © Questel "Orbit IPBI".)

5.2.5 Search strategy and search report

Unlike Derwent, Orbit presents a tool for automatically requesting a search report. In this topic, we will learn how to generate the document after performing a preplanned advanced search on the topic provided below. It is important to remember that the search proposal presented below is one of the innumerable possibilities that exist in a work of patentability analysis and related services. The purpose of this topic is to show the reader the management of *Search history* and the use of the operators and wildcard characters of the program. The results presented were generated in 2017, and therefore, the shown numbers below will present slight differences in the reader's search due to the frequent updating of the database. The search proposal is as follows:

Find some document about silica lipid nanoparticles containing polyethylene glycol and folic acid.

First, we searched for documents about silica nanoparticles in *Keywords* topic in *Advanced search*. It is important to mention that for all the cases below, the keywords were searched in the title, abstract, claims, description, object of invention, advantages over prior art drawback, and concepts. In this way, we typed the following expression in the search field: (*silica* **P** (*nanopartic+* **OR** *carr+* **OR** *vechicl+*)). The logic of this first search is to find documents whose synonyms of *nanoparticle* (carrier, vehicles) are in a paragraph similar to the word *silica*. When we click *Search*, the next page will cover the search results of that first set. The result in *Search history* is organized primarily as follows:

1) (*silica* **P** (*nanopartic+* **OR** *carr+* **OR** *vehicl+*)): 492,039 documents

After this step, we returned to the *Advanced search* page to search for documents related to lipids by typing the expression: (*lipid+* **OR** *c?olester+* **OR** *trig#cerid+* **OR** *fat+*). All terms were truncated with the wildcard character + in the suffix for the span of words in the singular and plural forms. The word *c?olester+* was truncated with the wildcard character ? in the letter "h" for comprising words *cholesterol* (and derivatives) and *colesterol* (and derivatives), the latter being a possible misspelling. We chose the words *cholesterol*, *triglycerides*, and *fatty acid* (*fat+*) as possible synonyms for the word *lipid*. *Search history* shows the following history:

2) (*lipid+* **OR** *c?olester+* **OR** *trigl#cerid+* **OR** *fat+*): 4,529,471 documents
1) (*silica* **P** (*nanopartic+* **OR** *carr+* **OR** *vehicl+*)): 492,039 documents

We have above searches 1 and 2 referring to documents about silica nanoparticles and lipids, respectively. To begin the search for silica lipid nanoparticles, we combined sets 1 and 2 with the **F** operator. To do this, we entered 1 **F** 2 in the *Combine strategies* field below the *Search history*, resulting in the following:

3) 1 **F** 2: 174,453 documents
2) (*lipid+* **OR** *c?olester+* **OR** *trig#cerid+* **OR** *fat+*): 4,529,471 documents
1) (*silica* **P** (*nanopartic+* **OR** *carr+* **OR** *vehicl+*)): 492,039 documents

Compare the number of documents covered in set 3 with the values in searches 1 and 2. We noted a refinement, whose patent quantity is still significant. Therefore, we attempted a new refinement of searches 1 and 2 using the more restricted **P** operator. To do this, we entered 1 **P** 2 in the *Combine strategies* field below the *Search history*, resulting in the following:

4) 1 **P** 2: 96,331 documents
3) 1 **F** 2: 174,453 documents
2) (*lipid+* **OR** *c?olester+* **OR** *trig#cerid+* **OR** *fat+*): 4,529,471 documents
1) (*silica* **P** (*nanopartic+* **OR** *carr+* **OR** *vehicl+*)): 492,039 documents

Note the refinement strategy hierarchically using the **F** and **P** operators. The reduction was from 174,453 to 96,331 documents. We could continue the line of reasoning using the **S** operator. However, since this operator still holds a very significant restraint profile, we maintained this strategy as "Plan B" and continue the process considering the other features of the proposed invention. In this context, the next step is to search for documents referring to polyethylene glycol by typing (*pol#et?#len+* **2W** *gl#c+*). This strategy is interesting because we used two distinct wildcard characters side by side in the word *polyethylene*. Initially thinking on the wildcard character #, the latter was used in the letters "y" to cover possible misspellings (poliethilene, polyethilene, poliethylene, etc.). The use of the wildcard character ? in the letter "h" of the word *polyethylene* covers misspellings as in *polyetylene* (none character covered by the wildcard character ?). This strategy comprises all possible misspellings mentioned above and their variants. The term *glycol* was also truncated with the # character in the letter "y" following the same line of reasoning. We know, finally, that the inserted wildcard character+covers words in the singular and plural forms. The **nW** operator used keeps the words *polyethylene glycol* in the order they are typed in the search field. The choice in the use of this joint operator is due to the fact that *glycol polyethylene* words (reverse order covered by the **nD** operator) are unusual and, if it occurs, may refer to distinct molecules or even to a descriptive part of a method. The spacing $n = 2$ was chosen to precisely encompass the combined terms *polyethylene glycol*. By typing (*pol#et?#len+* **2W** *gl#c+*) in *Advanced search*, we obtained the following result:

5) (*pol#et?#len+* **2W** *gl#c+*): 739,350 documents
4) 1 **P** 2: 96,331 documents
3) 1 **F** 2: 174,453 documents

2) (*lipid+* **OR** *c?olester+* **OR** *trig#cerid+* **OR** *fat+*): 4,529,471 documents
1) (*silica* **P** (*nanopartic+* **OR** *carr+* **OR** *vehicl+*)): 492,039 documents

The search 5 encompasses documents in which said polymer is cited. We continued the process in a sixth set referring to patents that mention folic acid by typing (*folic+* **2D** *acid+*) in *Advanced search*. Considering the likelihood of writing the expression *folic acid* also as *acid folic*, the **nD** operator was employed. The used wildcard character+includes the singular and plural forms of the words. The resulting search history is as follows:

6) (*folic+* **2D** *acid+*): 76,581 documents
5) (*pol#et?#len+* **2W** *gl#c+*): 739,350 documents
4) 1 **P** 2: 96,331 documents
3) 1 **F** 2: 174,453 documents
2) (*lipid+* **OR** *c?olester+* **OR** *trig#cerid+* **OR** *fat+*): 4,529,471 documents
1) (*silica* **P** (*nanopartic+* **OR** *carr+* **OR** *vehicl+*)): 492,039 documents

Now that we have searches 5 and 6 related to documents quoting polyethylene glycol and folic acid, respectively, and we will cover the following strategies (a) and (b) described below:

a. We will combine the documents about polyethylene glycol and folic acid using the **OR** operator and then will combine it with the search about silica lipid nanoparticles to obtain patents about silica lipid nanoparticles containing polyethylene glycol or folic acid—it is worth remembering that the search proposal is to find some patent document about silica lipid nanoparticles containing polyethylene glycol and folic acid simultaneously. If the search was performed using the **AND** operator or some joint operator between the searches about polyethylene glycol and folic acid, there is the possibility that no document could be found, and on that occasion, we would return the process to a wider search profile. The achievement of advanced search in a broader way following a gradual refinement process allows the analyst an overview of the characteristics of invention that already exist. This information will also be useful in developing the patent protection strategy (in case of novelty) since the analyst is aware of the limits by which the invention can be protected and has knowledge of the aspects already covered in the state of the art;
b. We will combine the documents about polyethylene glycol and folic acid using joint operators and then combine it with the search about silica lipid nanoparticles to obtain patents about silica lipid nanoparticles containing polyethylene glycol and folic acid, if strategy (a) returns significant amount of documents.

In the first aspect, we combined searches 5 and 6 with the **OR** operator by typing (5 **OR** 6) in the *Combine strategies* field below the history of *Search history*. This result is part of strategy a. Next, we will combine searches 5 and 6 in two ways: using the **P** operator to search for the words *polyethylene glycol* and *folic acid* in the same paragraph by typing (5 **P** 6) in the *Combine strategies* field, that is, both simultaneously present in the found documents and using the **S** operator to search for these words in the same sentence by typing (5 **S** 6), that is, also simultaneously present in the found documents. Both results will be part of strategy b and correspond to a search refinement reasoning within the same strategy, if necessary. *Search history* will return the following results:

9) 5 **S** 6: 1,136 documents > for refinement of strategy b, if necessary
8) 5 **P** 6: 4,798 documents > for the strategy b
7) 5 **OR** 6: 786,673 documents > for strategy a
6) (*folic*+ **2D** *acid*+): 76,581 documents
5) (*pol#et?#len*+ **2W** *gl#c*+): 739,350 documents
4) 1 **P** 2: 96,331 documents
3) 1 **F** 2: 174,453 documents
2) (*lipid*+ **OR** *c?olester*+ **OR** *trig#cerid*+ **OR** *fat*+): 4,529,471 documents
1) (*silica* **P** (*nanopartic*+ **OR** *carr*+ **OR** *vehicl*+)): 492,039 documents

Following the strategy a, we will search for documents about silica lipid nanoparticles containing polyethylene glycol or folic acid combining the searches 4 (silica lipid nanoparticles) and 7 (polyethylene glycol or folic acid) using the **P** operator by typing (4 **P** 7) in the *Combine strategies* field, resulting in the following:

10) 4 **P** 7: 26,826 documents > strategy a
9) 5 **S** 6: 1,136 documents > for refinement of strategy b, if necessary
8) 5 **P** 6: 4,798 documents > for the strategy b
7) 5 **OR** 6: 786,673 documents
6) (*folic*+ **2D** *acid*+): 76,581 documents
5) (*pol#et?#len*+ **2W** *gl#c*+): 739,350 documents
4) 1 **P** 2: 96,331 documents
3) 1 **F** 2: 174,453 documents
2) (*lipid*+ **OR** *c?olester*+ **OR** *trig#cerid*+ **OR** *fat*+): 4,529,471 documents
1) (*silica* **P** (*nanopartic*+ **OR** *carr*+ **OR** *vehicl*+)): 492,039 documents

A significant number of documents were observed. Thus, we followed the strategy b by combining search 4 with search 8 to find patents about silica lipid nanoparticles containing polyethylene glycol and folic acid (proposal of this case study) using the **P** operator. We then typed (4 **P** 8), resulting in the following:

11) 4 **P** 8: 552 documents > strategy b

10) 4 **P** 7: 26,826 documents
9) 5 **S** 6: 1,136 documents > for refinement of strategy b, if necessary
8) 5 **P** 6: 4,798 documents
7) 5 **OR** 6: 786,673 documents
6) (*folic*+ **2D** *acid*+): 76,581 documents
5) (*pol#et?#len*+ **2W** *gl#c*+): 739,350 documents
4) 1 **P** 2: 96,331 documents
3) 1 **F** 2: 174,453 documents
2) (*lipid*+ **OR** *c?olester*+ **OR** *trig#cerid*+ **OR** *fat*+): 4,529,471 documents
1) (*silica* **P** (*nanopartic*+ **OR** *carr*+ **OR** *vehicl*+)): 492,039 documents

There is a significant refinement for 552 documents, still broadly from the analytical point of view. Fortunately, we still have the option of refining this result using search 9 as a tool. To do this, we typed (4 **P** 9) to obtain the following:

12) 4 **P** 9: 9 documents > strategy b refined with search 9
11) 4 **P** 8: 552 documents
10) 4 **P** 7: 26,826 documents
9) 5 **S** 6: 1,136 documents > for refinement of strategy b, if necessary
8) 5 **P** 6: 4,798 documents
7) 5 **OR** 6: 786,673 documents
6) (*folic*+ **2D** *acid*+): 76,581 documents
5) (*pol#et?#len*+ **2W** *gl#c*+): 739,350 documents
4) 1 **P** 2: 96,331 documents
3) 1 **F** 2: 174,453 documents
2) (*lipid*+ **OR** *c?olester*+ **OR** *trig#cerid*+ **OR** *fat*+): 4,529,471 documents
1) (*silica* **P** (*nanopartic*+ **OR** *carr*+ **OR** *vehicl*+)): 492,039 documents

We found a group of nine documents that can be easily analyzed. The summary of the above *Search history* can be observed in Figure 5.10.

By clicking *Show results* in search 12 (final result), we will find a relevant patent titled *Magnetic resonance imaging contrast agents containing water-soluble nanoparticles of manganese oxide or manganese metal oxide* whose patent number is WO2007064174. By opening the patent description page by the orbit page or via PDF document, we will appreciate that the claims of that document protect any of the inventions including MnO_2 nanoparticles with folic acid, MnO_2 nanoparticles with polyethylene glycol, MnO_2 nanoparticles with inorganic support based on silica, and lipid MnO_2 nanoparticles. No protection was found comprising silica lipid nanoparticles with polyethylene glycol and folic acid. Therefore, the proposed invention in this case study could be patented with respect to the protection limits observed in the search history shown in Figure 5.10.

As we have studied, the search strategy can be fully saved by clicking *Search history* and *Save entire strategy* and following the brief steps for

Figure 5.10 Panel of the *Search history* of the case study through Orbit. (Source: © Questel "Orbit IPBI".)

storing information in *My saved searches* topic. A differential feature of Orbit platform is the generation of a search report (in word, PDF, etc.). For this purpose, we click *Show results* from search 12 (final search) to view the respective result page. Next, click the *Select* icon below where it reads *9 results for 4 P 9—Collection: FAMPAT* and select the option *All records* to select all found documents. Once it is done, click the *square-like with an arrow down* icon and select the file extension in which the report will be saved (PDF, TXT, XLS, XML, etc.). In our example, we chose to save the report as PDF.

A window will appear asking for the items that we want to include in the report. In this example, we chose the document template as *First page like*. The items we chose to be contained in the report are the first image of the invention and the search strategy (*Search history*) by selecting *First page like* and *Strategy (complete)*, respectively. The analyst has the option to include the remaining items, such as description of each invention found, claims, all images, key info, legal status, cited patents, and nonpatent literature, as noted in Figure 5.11.

By clicking *Next*, a final window will open requesting the option of extracting the document via download by Orbit, zipped file, or e-mail. In this example, we selected the option to obtain the file via download by selecting it and by clicking *Finish*. Once it is done, the file will be uploaded, and the link *Click here to download file* will be available to obtain the report.

This chapter closes the concepts of advanced patent search through two important platforms, Derwent and Orbit. We have seen that both own tools for accurate results when compared to the basic searches. In the course of this chapter, the importance of the knowledge regarding advanced search for professionals from patent market, academic

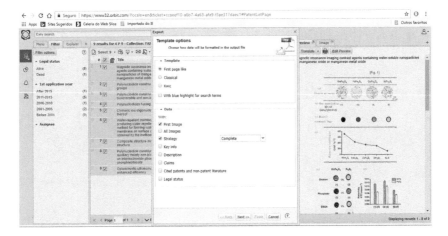

Figure 5.11 Exporting a search report by Orbit. (Source: © Questel "Orbit IPBI".)

researchers, and research and development professionals is evident since
the mastery of the studied techniques allows analyzing the possibility
of developing a novel invention based on preestablished ideas; knowing
the characteristics of a particular invention based on the obtained results;
planning improvements; knowing whether an object is protected by patent; and enabling a complementary market analysis in the search for
potential suppliers, competitors, and partner companies. From the point
of view of intellectual property in the context of patents and utility models, advanced search enables accurate analysis of patentability and supports the process of designing product or process protection by patent.

chapter six

Practical exercises by the advanced patent search

6.1 Exercises

1. You are an analyst from a technology innovation team from a company focused on research and development. A novel nanomaterial with high market and technological potential was developed and your team have been requested for performing a patentability analysis. This product is described below:

Graphene oxide functionalized with carbon dot containing the drug doxorubicin for antitumor applications

For the search you have the following keywords:
Graphene, carbon point, doxorubicin, and tumor cell

Using only Derwent advanced search for the problem, you are asked to
   a. search for other similar keywords to complement those provided above (acronyms, molecular formulas, synonyms, etc.) using Google web.
   b. elaborate a search strategy explaining your reasoning.
   c. present the performed search showing the involved steps and related results—consider for the search the Boolean and joint operators as well as the wildcard characters.
   d. present the most relevant patent and
      – show and explain similar features between the found technology and the novel product of your company, presented above;
      – show and explain different features between the found technology and the novel product of your company, presented above.

Important: For a quick and accurate analysis, investigate the claims of found patents, once it provides the protect features of the technology and shows the limits of protection.
2. The CEO of your Research and Development company will take the decision to act in a promising and new niche market: Tissue

Engineering. In this way, the leader has been had an idea about developing a new product, and he wants to know whether it is novel compared to the state of the art. The aim, therefore, is to investigate the novelty of the proposed idea, and your team were requested to do this work. Below it follows the product features the CEO intendeds to develop:

A nanocomposite based on hydroxyapatite, polyurethane, and poly(vinyl alcohol) for bone graft.

Important information:
Hydroxyapatite: porous material present in bones, with high mechanical strength (geometry and structure of the material);
Polyurethane: polymer used as a matrix for tissue formation (not bioactive—used to improve mechanical properties and bioactivity);
Poly(vinyl alcohol): biodegradable and biocompatible polymer (promotes elasticity to the material and greater dispersing capacity).

The broadest keywords are as follows:
Polyurethane, hydroxyapatite, polyvinyl alcohol, bone, tissue, and collagen
  Your company holds only Orbit as an advanced patent search program, and therefore, this will be your only tool for analysis. Based on the above information, you are requested to
a. search for other similar keywords to complement those provided above (acronyms, molecular formulas, synonyms, etc.) using Google web.
b. elaborate a search strategy explaining your reasoning.
c. present the performed search showing the involved steps and related results—consider for the search the Boolean and joint operators as well as the wildcard characters.
d. present the most relevant patent and
    – show and explain similar features between the found technology and the novel product of your company, presented above;
    – show and explain different features between the found technology and the novel product of your company, presented above.

Important: For a quick and accurate analysis, investigate the claims of found patents, once it provides the protect features of the technology and shows the limits of protection.
3. You are an academic researcher with extensive knowledge in advanced patent search and the university in which you work holds

as search tools the Derwent and Orbit platforms. You need to plan a research project whose information you need to run it is contained in a patent document whose patent number, title, or inventors/assignee names is not remembered. You know, however, that this specific patent mentions, as one of the objects of the invention, a ferromagnetic nanoparticle containing nitric oxide groups (also known as S-nitrosothiol groups or SNOs) on the outer surface, for biomedical applications. One of your graduate students has saved the figures of the said patent document and has shown you as a search support (Figures 6.1 and 6.2).

Important information:
NO or S-nitrosothiol groups: S-nitrosothiol groups are biologically active molecules that act in processes of cutaneous healing, and vasodilatation of blood vessels, among others.
Ferromagnetic component: It is important for identification of nanoparticles in cellular internalization by nuclear magnetic resonance imaging.

With the above information, we ask to
a.   develop a search strategy to find the unknown patent.
b.   with the elaborated search strategy in item (a), execute it in Orbit and present the results.

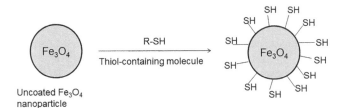

Figure 6.1 Example of an image that might be found at patent platform search system.

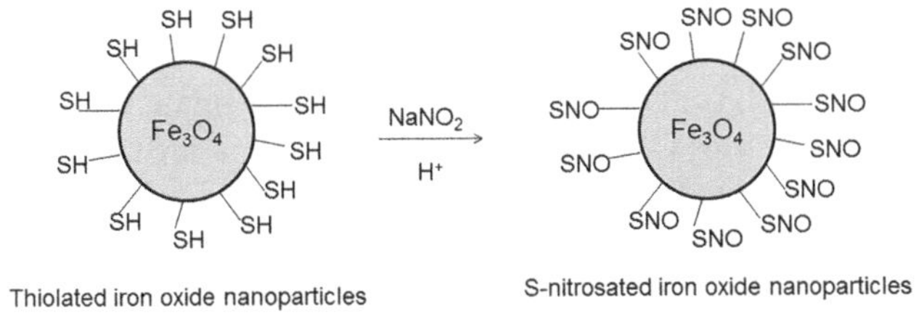

Figure 6.2 Example of an image that might be found at patent platform search system.

c. with the obtained results in the search for item (b), present the patent number, the title of the patent, the inventors, and the involved assignees of the unknown patent.

d. with the search strategy elaborated in item (a), execute it in Derwent and present the results.

e. compare the final results of Orbit and Derwent and comment about similarities and/or differences in obtained results on both platforms.

6.2   *Expected answers*

1a. Some keyword options with synonyms are shown below:

For graphene, we have the following options:
Graphene **OR** graphene oxide
(we call these keywords as "graphene terms")

For carbon dots, we have the following options:
carbon dot **OR** carbon quantum dots **OR** carbon nanoparticles **OR** CQDs **OR** CDs **OR** cdots **OR** small carbon nanoparticles
(we call these keywords as group "carbon dot terms")

For doxorubicin, we have the following options:
doxorubicin **OR** adriamycin **OR** C27H29NO11 **OR** DOX
(we call these keywords as group "doxorubicin terms")

For the expression tumor cell, we have the following options:
tumor cell **OR** cancer **OR** carcinoma **OR** tumor
(we call these keywords as group "tumor cell terms")

1b. In the first aspect, we will open four sets referring to the above keyword groups (graphene, carbon dots, doxorubicin, and tumor cell). First, following a broad strategy, we will combine all searches with the **AND** operator. If the number of documents is relevant, we will combine the keyword groups about "graphene terms" and "carbon dot terms" using the NEAR/x operator (beginning with $x=3$). When two nanostructures are combined, their respective terms are commonly written close to each other in documents. Next, we will combine this last generated search with that related to "doxorubicin terms" using the **AND** operator. The mention of drug addition may occur near or far from terms referring to the nanostructures. If the search shows a significant number of documents, we will refine this by combining it with the search about "tumor cell terms" using the **AND** operator, for the same reason explained above. If the numbers are yet significant, the terms graphene-carbon dots (before combined with NEAR/3 operator) will be combined with the "doxorubicin terms" using the NEAR/5 operator. From here, a refinement strategy will be through the use of the NEAR/4 and NEAR/3 operators. We expect this refinement to be enough to find few relevant documents for analysis.

1c. The search results are shown in the *Search history* (Figure 6.3).

Sets 1–4 refer to the four groups of keywords we initially planned. Note the use of wildcard character * for singular and plural forms of words and the use of wildcard character $ substituted for the letter "y" on the word adriamycin. **NEAR/2** operators were also used in expressions with at least two keywords as in *tumor cells* and *small carbon nanoparticles*. Interestingly, a combination of the four mentioned searches using the **AND** operator resulted in a set of only six documents to be analyzed without the need to focus on the refinement strategy initially thought.

1d. The most relevant found document owns the title *The therapeutic composition useful for targeting tumor cells contains a set of nanovectors, one or more active agents, one or more active agent enhancers and one or more targeting agents* and patent number as US2015216975. The covered technology is the material based on graphene quantum dots containing doxorubicin:

- The found technology and the proposed invention are similar due to the fact of comprising materials based on graphene and doxorubicin and quantum dot properties.

Figure 6.3 Search history. (Source: Web of Science™ images are reproduced with permission from Clarivate Analytics © Clarivate Analytics 2018. Web of Science™, InCites™, Journal Citation Reports™, Essential Science Indicators™, Endnote™, Publons™, Clarivate Analytics™, and the logo of Clarivate Analytics are trademarks of Clarivate Analytics and its affiliated companies and are shown with permission. All rights reserved.)

- The differences between the found technology and the proposed invention lie in the nature of the nanostructure. In the found technology, graphene is the own quantum dot, explaining why it is designated as graphene quantum dots. In the proposed invention, we have a nanostructure based on graphene oxide combined with another nanostructure based on carbon dots (two different materials). Furthermore, the mentioned technology and invention refer, respectively, to graphene and graphene oxide, two chemically different nanomaterials.

2a. The provided keywords were organized by groups of synonyms, acronyms, and/or molecular formulas below. These terms were found using the Google web platform:

"Hydroxyapatite terms"
(*hydroxyapatite* **OR** *hydroxylapatite* **OR** *Ca5PO43OH* **OR** *Ca10PO46OH2* **OR** HA)

"Polyurethane terms"
(*polyurethane* **OR** *PUR* **OR** *PU carbamate polymer*)

"Poly(vinyl alcohol) terms"
(*poly vinyl alcohol* **OR** *PVOH* **OR** *PVA* **OR** *PVAI* **OR** *C2H4O polymer*)

"Application terms"
(*bone* **OR** *tissue* **OR** *collagen*)

2b. First, we will perform four sets related to the above four keyword groups related, respectively, to "Hydroxyapatite terms," "Polyurethane terms," "Poly(vinyl alcohol) terms," and "Application terms." After this step, we will search for documents related to the hydroxyapatite material with polyurethane and poly(vinyl alcohol), that is, we will try to find patents that mention the product itself regardless of its final application. To do this, we first will combine the searches using the **AND** operator. In the possibility of numerous found documents, we will refine the mentioned sets with the **5D, 4D,** and **3D** operators, in that order, within the need of refinement. We would then have a set of documents that mention such material, regardless of its application. Within the need for more refinement, the resulting search will be combined with that referring to the "Application terms" using the **AND** operator, followed by a refinement reasoning with the operators **F, P,** and **S** operators, in that order. If in the search in which the **S** operator used covers many documents, we will restart another refinement process. For this, searches with these last three operators will be ignored, and the search for the "Application terms" will again be combined with the one that mentions the product itself, using the **8D**, **7D**, and **6D** operators, in that order—in this case, we will use $n=6$ as the limit since application-related keywords are often found to be far from the terms related to the material itself within the document. Important: The process will be configured to search the keywords in the title, abstract, claims, description, object of invention, advantages over prior art drawbacks, and concepts of the documents.

2c. The performed search in Orbit presented the *Search history* (Figure 6.4):
   We found ten documents for analysis as observed in Search 7. Characters+were used to comprise words in the singular and plural forms, while wildcards characters # were used in the letters "y" of words such as *polyurethane* for spelling errors. Interestingly, the word *hydroxyapatite* was truncated as h#drox#?apatit+. The inserted character ? is important to find the words *hydroxyapatite* and *hydroxylapatite* that are found since the letter "l" can be included in the one character option and be absent in the option of none character, according to the rule of the character ? itself. In the search regarding the material itself using the **AND** operator, the result comprises numerous documents. When we refined the latter with the **5D** operator, the final result was only 12 documents. We then combined the latter with the search referring to the "Application terms" comprised in Search 4, resulting in ten documents.

2d. The most relevant found document is the patent BR102013014155, titled "Bionanocomposite for bone recovery." This technology

Figure 6.4 Search history. (Source: © Questel "Orbit IPBI".)

exactly comprises a hydroxyapatite-based material with polyurethane and poly(vinyl alcohol) for tissue engineering, being wholly
similar to the invention envisioned by the company's CEO. Here, we
have a case of an idea already existing in the state of the art.

3a. In the first aspect, a search will be carried out taking into consideration the documents referring to magnetic nanoparticles. Next,
a set of patents mentioning nitric oxide, s-nitrosothiol groups, and
derivatives will be performed. Both searches will first be combined
using the **AND** operator and, if necessary, will be refined using joint
operators from the widest to the narrowest according to tools available in Orbit and Derwent. In this exercise, the reader is expected
to have autonomy to predict and find out keywords on the Internet
without previous tips given in the problem. In fact, the reading of
the problem itself provides key information to instigate reasoning
about possible keywords, synonyms, acronyms, molecular formulas,
and so on.

3b. The results obtained on the Orbit platform are shown in Figure 6.5.
A total of 16 patents were found for analysis.

3c. The unknown patent sought is the document BR102012016127 whose
title is *Functionalized ferromagnetic nanoparticles, functionalization process of ferromagnetic nanoparticles, use of the outputs and pharmaceutical composition.* The inventors are Marcelo Ganzarolli de Oliveira,
Amedea Barozzi Seabra, Paula Haddad, and Rosângela Itri. Assignee
of the patent is the State University of Campinas.

3d. The obtained results on Derwent platform are shown in Figure 6.6.

Figure 6.5 Orbit platform. (Source: © Questel "Orbit IPBI".)

Figure 6.6 Derwent platform. (Source: Web of Science™ images are reproduced with permission from Clarivate Analytics © Clarivate Analytics 2018. Web of Science™, InCites™, Journal Citation Reports™, Essential Science Indicators™, Endnote™, Publons™, Clarivate Analytics™, and the logo of Clarivate Analytics are trademarks of Clarivate Analytics and its affiliated companies and are shown with permission. All rights reserved.)

The search resulted in 12 patent documents. Interestingly, using the same used strategy for Orbit search, the patent BR102012016127 was not found in this search.

3e. To compare the results, one should analyze the found documents in common for both platforms and identify those that are unique to each one. For this, the patent number information of all found documents in Derwent search was entered into the topic *Numbers* of Orbit, to be opened as a set comprising the covered patents in the final result of Derwent. The keywords have been inserted as (CN106568936 **OR** CN106010500 **OR** CN105259243 **OR** CN103041791 **OR** CN101647780 **OR** US2010111844 **OR** US2010111843 **OR** US2010111842 **OR** US2010111845 **OR** CN101551390 **OR** CN101551391 **OR** WO2008131528). Note that all these typed numbers correspond to the 12 found documents in Derwent, generating the Set 7 in *Search history* of Orbit. We then have sets 6 and 7 for the final search results for Orbit and Derwent, respectively. We subtracted from search 6 (with the **NOT** operator) all documents from search 7 by typing (6 **NOT** 7). A total of 16 documents were also generated, that is, there was no subtraction effect. This shows that the 16 found documents in Orbit are different from the 12 found documents in Derwent. To

Figure 6.7 Comparison between Derwent and Orbit. (Source: © Questel "Orbit IPBI".)

complement this previous conclusion, we typed (6 **OR** 7), and the number of documents was 28, equal to the sum of 16+12. The comparative results are shown in Figure 6.7.

The above results show the complementary of the two advanced search platforms. This exercise reinforces the importance of using at least two search platforms in the process of analyzing novelty in the patent system. Note that the search strategy was similar in both cases.

chapter seven

How to read a patent

7.1   *To know the meaning of the information from the first page of the patent*

The front page of patents is very useful since it has all the information of the full process of patent registration (see also page 11). Then, it is necessary to know exactly the meaning of any part of this page and see the sequence of numbers:

Patent number (1): The patent is assigned a unique sequentially generated number from the United States Patent and Trademark Office (USPTO; in this case in US) that is used to identify the patent. Patent attorneys, for simplicity, use the last three digits of the patent number to refer to a patent. In this case, the patent might be referred to as "the 628 patent."

Publication data (2): This is important for priority data of the patent registration. This could be useful in legal problems with another similar or identical patent. Whoever publishes first is the owner of the patent.

Inventors (3) and (5): A patent always has at least one inventor, who is a person and not a company. The first inventor is indicated in (3). Patent attorneys use this name to refer to a certain patent listed in (3), as with the Nunes patent. The inventors in a patent are listed in (5) with complete address and country of residence. It is interesting to comment that, in the case of multiple inventors, each inventor can commercialize independently without any approval of the other inventors, unless the inventors have signed an agreement to the contrary. More importantly, if an applicant does not faithfully supply all of the actual inventors' names, and the applicant does so with misleading intent, the patent probably will be found to be invalid and inapplicable. It is also possible for an inventor who has been eliminated from a patent to take legal action and be added to the patent after the patent has been published.

Title (4): The title must indicate the subject of the patent. But we have to be careful, with the title, since in many cases, it is misleading. Therefore, do not judge a patent based only on the title.

Assignee (6): This section is important since it gives an indication that someone other than the inventor(s) ensures rights in the patent.

In this specific case, the assignee is also an inventor. Rights to a patent are often transferred by the inventors to another person or entity before the patent issues. In general, in a company, most employers require their employees to assign any inventions they make within the scope of their contracts to the employer.

Application number (7): This number is sometime called the application serial number and is useful for searching related patents. This number is also useful to identify the application before being accepted in the patent office.

Filing date (8): This date is extremely important information in a patent. The date or the issue date will indicate the period of the patent that will be executable (in general, for utility patents, it is 20 years for many countries).

This date of a patent may also be used as references to determine the patentability of the invention and also to determine the date at which the patent is effective as a reference against other applications.

Patent classification and field of search (9): The patent classification is used to group patents with similar technologies. The field of search indicates the classifications in which the patent examiner searched to locate prior art. These data can be helpful when performing a patent search for related inventions.

Abstract (10): The abstract gives a brief overview or summary of the invention. This information is used by patent searchers and examiners as an initial point when determining whether a prior patent is relevant to an application being considered. Here, it is important to be careful since just as the title can be misleading, so can the abstract. Hence, it is important that we do not rely solely on the abstract to determine the scope of a patent (Note: for more detailed information to these topics, see Burn (2014); see also: www.bios.net/daisy/bios/204/version/live/part/4/data).

7.2   What is the most important to read a patent quickly?

7.2.1   Important patent parts

Now, in the case that you are interested in reading a patent for business, academic, or industrial reasons, then you have to start with the important parts of a patent. Patents have three major parts (and a bunch of bibliographic and classification data to help organize them), namely, abstract, claims, and specification (Way Better Patents™, http://www.waybetterpatents.com/how_to_read_a_patent.html).

7.2.1.1   Abstract

The *Abstract* is the summary of the invention indicating the technical area to which the invention belongs. It is expected to be drafted in such a way that it permits easy comprehension of the technical problem, the nature of the solution of the specific problem through the invention, and the important use or uses of the invention. The abstract is supposed to be short (e.g., 50–150 words long). This part of the patent represents the patent market's statement: the object (the thing), how it does, and what problem does it solve.

7.2.1.2   Claims

The *Claims* are what the inventor possesses. The claims are exactly what the inventor invented and what is actually the patent is protecting. In other words, it is the inventor's exclusive rights in the patent.

A claim is only one sentence, starting with a capital letter and ending with a period. The inventor refines his or her invention down to that one sentence. Therefore, if the patent has 20 claims, then the patent covers 20 inventions, and in general, the inventors try to describe the most important claim first.

After reading claim 1, search around and have a look at the other claims to understand better what they describe. The claims come in two ways: independent and dependent. If the claim has the phrase such as "The system comprise ...," it is an independent claim. If a claim has a phrase such as "The system as claimed in Claim 1," it is a dependent claim. Therefore, it is necessary to read both claims together to understand the full scope of the invention being claimed.

7.2.1.3   Specification

The *Specification* is supposed to teach someone, a person ordinarily skilled in the art, how to make or practice the invention.

In general, the specification starts with a significant group of paragraphs: background and summary of the invention and list of figures.

The background of the invention describes the actual knowledge at the time the inventor reported his or her invention. This part of the patent gives helpful context on what the invention is about, the state of the art which is relative to the invention, and an account of the problems that the invention will solve. The aspects of what really the patent is all about or benefit or improvement is given by the summary of the invention. The summary of the invention can also help us to understand what the patent is about before you have any profound analysis of the patent. The next is the list of figures that the inventors use to explain the invention. Of course, if the list of figures is well done, it is possible to extract a good source of information related to that patent.

7.2.2   Reading the patent

General comments:

To understand the legal effect of a patent, in other words, what is invented, what is forbidden, we should read its claims.

On the contrary, if you need to understand the patent's "teachings" or effect as prior art, you have to read the figures and written description. However, for legal effect, read the title and abstract, scan the figures, and read the first claim.

If the claim is incomprehensible, start reading the written description, using the figures as a guide. It is necessary to remember that every word of the claim matters and that the more words, the narrower the claim.

7.2.2.1   Title and the abstract

General comments:

As discussed previously, the title must indicate the subject of the patent. But we have to be careful with the title, since in many cases, it is misleading. Therefore, do not judge a patent based solely on the title. The title is a short text describing the contents of the application. In the title field, you can enter up to ten search terms, separated by spaces or the appropriate operators. However, titles can be intentionally ambiguous. It is, therefore, recommended that you enter your search terms in the title or abstract field or to search in the description and claims.

What we expect from the abstract? The abstract should describe the invention neatly in as few words as possible. It is found that the patent agent could use a reduced version of the first paragraph of the summary as the actual abstract. In general, the abstract is not really considered by the patent analyst for these reasons. Please do not think that the abstract should be deceptive or poorly written. One possible danger with abstracts is that they may disclose some patentable aspects of the invention not found in the description.

In other words, it is necessary to be careful not to conclude something only from the abstract (WIPO, 2007a,b).

7.2.2.2   Description

This section, usually described as the "preferred embodiment of invention" section or sometimes as the "disclosed embodiment of the invention" section, imbues life into the claims and gives an adequate interpretation of the invention so that a non-skilled person has a the chance to perform and understand the invention. A particular description section must be closely correlated to the drawings. It is important to know that this section cannot be redressed once the application has been filed. Thus, the inventor or the patent agent should take care that the patent application clearly reflects the disclosure material fixed by the inventors and gives enough information to enable a common artisan to reproduce the invention (WIPO, 2007a,b).

7.2.2.3   List of figures

The patent must be prepared with clear visual supporting materials that describe the invention. In some cases, patent laws require that every claimed statement be shown in a drawing. In some cases, when the drawing is quite well represented in detail, reading the detailed description section simply corroborates in words the information given in the drawings. Of course, this will not be possible with all inventions.

7.2.2.4   Claims

The first thing that the inventor must prepare is the claims for the invention. This will often provide confirmation to the inventor that anyone can understand the invention. Remember that the claims are the important part of the legally operative section of a patent application; everything is decided around the claims. It is generally preferable to draft the claims first, but some conditions may not grant the patent agent with this luxury. As an example, assume that a patent agent receives a technical paper from an inventor who tells the patent agent that the patent application needs to be filed immediately because of an oncoming public disclosure of the invention. Because of the critical importance of claims, sometimes it is necessary to carefully revisit the specifications to check that the object of the patent was well understood.

In a patent, the claims generally come last, and the claims are usually preceded by the phrase "What is claimed is …" or "I claim …." Many patent agents believe that claims are the heart of a patent, which defines the border of exactly what the patent does, and does not, cover. In other words, the inventor has the right to exclude others from making, using, or selling the things which are described exactly by the claims. To understand the claims of a patent is the key to defining whether a given product or process violates the patent.

The reader must be aware that this is not an easy thing to do because the claims are written in a legalistic and inflated way. Be careful with the claims since they do not exactly mean what they say. In that case, it is also necessary to read the specification and drawings (see www.bpmlegal. com/howtopat.html).

7.2.2.4.1 Dependent claims   If any claim that initiates with "The xxx of claim (x)" is fundamentally a refinement or detail with narrower scope than the parent claim, it is a dependent claim.

Example: Patent: US2006/0093628: "A formulation in injectable form for the protein aggregate of ammonium and magnesium phospholinoleate-palmitoleate anhydride, <u>according to claim 1</u>, obtained by ...." It is a dependent claim.

7.2.2.4.2   *Independent claims*   Any claim that starts with "The xxxx comprising or in which or wherein ...." It is an independent claim.

Example: Patent: US2006/0093628: "We claim: 1. An immunomodulator <u>comprising</u> a protein aggregate of ammonium and magnesium phospholinoleate-palmitoleate anhydride, which includes the presence of ...." It is an independent claim.

The claims are the only part of the patent that has any actual legal applicability.

7.3   Special case: procedure for a rapid analysis of a patent

To quickly and superficially read a patent, or if a nonspecialist wants to know about the subject of the patent and get an idea of what is or is not important to his or her interests, it is possible to follow this process:

7.3.1   Ignore the title, drawing, abstract, and specification

In general, not all the patents, but, the title of the patent can be just whatever the author wants. They may describe the product or process being improved, but not the new invention. Patent drawing is generally very difficult to read, and sometime indirectly influences the applicability of the patent. In an article, the abstract is the most important part since it is a short, direct summary of the important point of the article. But in many cases, in patents, this is not the case. In your first reading of the specifications of the patent, the data presented in the abstract, such as background or state of the art of the patent, is not necessary to the reader.

7.3.2   Ignore the dependent claims

Since the dependent claim is a refinement of a parent claim, it exhibits, in general, a narrow scope of the patent, for a rapid reading of the patent is not necessary to analyze them.

7.3.3   Most importantly, read the independent claims

The claims are the most important part of the patent since they have the actual legal applicability. Although they are not easy to read for a nonspecialist, they are short and pointed out the right direction. This possibility is interesting since if you need a quick summary of a patent, just directly read the independent claims and you will have a rapid comprehension of the subject of the patent (see more details in these suggestions on www. danshapiro.com/blog/2010/09/how-to-read-a-patent-in-60-second).

7.3.4 *Another alternative is the following*

In a case of the researcher need to read many patents and the time is scarce, the logical procedure is first of all to read the first page as discussed previously. Filing date and grant data led us to determine the period of validity of the patent. The title and the abstract of the patent led us to have, although superficial, what is the subject matter. The independent claims of the patent give us at least some idea of the scope of the matter. After this quick reading, you will have a basic understanding of the patent (Burn, 2014).

chapter eight

How to write a patent
Basic information

8.1   Introduction

Writing a patent requires different languages and structures compared to scientific publications. The key points you should keep in mind in writing a patent are as follows:

i. You have to update the state of the art and highlight the novelty, clearly demonstrating that you have an invention;
ii. Patents have financial purposes, and therefore, it is important to demonstrate the advantages of your inventions, how it might improve the actual state of the art in the appropriate subject, and how it would contribute the technological development;
iii. Claims are the heart of your patent. Special attention should be given in writing the claims;
iv. Try not to be so restricted;
v. Your invention application must be sufficiently described to prove that you have a novelty at the date of patent filing.

8.2   Parts of a patent application

This chapter presents and discusses how to write each item of a patent application, giving some advices and tips.

The parts of a patent application are:

i. Title
ii. Inventors
iii. Abstract
iv. Claims
v. Drawings
vi. Background of the invention
vii. Summary of the invention
viii. Brief description of the drawings
ix. Detailed description of the embodiments (examples).

8.2.1   Title of the invention

The title of invention, which is placed at the top of the first page, should be technically accurate and succinct and should describe the invention. Avoid long titles since there is a limit of words. Suitable titles have 2–7 words. The title should reflect on the principle of the invention, and it may clarify the claims. Laudatory statements, such as "high-performance process" and "improved biocompatible nanoparticles," may be part of the title in a chemical invention. Avoid narrow title. Examples of titles in a chemical invention are as follows: "Methods for producing high purity gold nanoparticles," "Stabilized iron oxide nanoparticles and their applications," and "Nitric oxide delivery smart materials."

8.2.2   Inventors

It is necessary to indicate all the inventors involved in a patent. The inventors in a patent are listed with complete address and country of residence.

8.2.3   Abstract

The abstract of an invention should contain a maximum of 150 words (usually 50–150 words), in a single paragraph, and it begins on a new page. It is expected that the abstract presents the novelty but not discusses the merits of the invention. The abstract should not compare the invention with the prior state of the art. In other words, you do not need to prove that you have a novelty in the abstract but only present the novelty. In this sense, in a chemical invention, the abstract of the patent that describes a new method to synthesize a material should describe how it is made; it should contain the steps of the process. A patent of a new chemical compound should present its identity and nature, as well as its uses. It is not expected that the abstract proves that this compound is new compared to the state of the art. Be aware to avoid the description of some invention features in the abstract that are not found in the patent specification. Avoid writing the abstract before writing the claims and the detailed description of the invention. It is preferred to write the claims and the detailed description of the invention before writing the abstract. When writing the abstract, keep in mind the following question: "Are the features described in the abstract disclosed in the specification?" It must be clear whether the answer is strongly "no"; hence you should rewrite the abstract or modify the specification.

8.2.4   Claims

Claims are the heat of an invention, the most important item. Special attention should be given for writing the claims. A patent application must contain one or more claims. It begins on a separate page and is numbered, and usually, it begins with the following sentence: "What is claimed." Generally, the claims are presented in order from the broadest to the narrowest. Claims are enumerated and grouped. If a patent contains only one claim, it need not be enumerated. As the claims are the most important section of a patent invention, a high level of accuracy is necessary in selecting appropriate words and expressions. It should be noted that intellectual property attorneys should attempt to write the claims, after reading and discussing with the inventors the detailed description of the invention and the novelty. However, the inventors should always try to propose the claims in a concise fashion. If the claims are not written correctly, it could be invalid. Claims define the invention and distinguish it from the state of the art. The aim of the claim is to describe the scope of the invention. For example, in a chemical invention, claims describe a compound that is obtained from several reactions; however, the claims should not teach how to prepare this compound. Detailed description of the invention and examples describes how to make the chemical compound. Each claim needs to have a specific purpose and meaning, in accordance with the invention principle and uses. When preparing the claims, you should start with broader claims and then to narrow claims. Broader claims can be considered obvious; therefore, it is necessary to include narrow claims to specify the general claim. The requirements for a claim are as follows: (i) it must be useful, (ii) it must be new, which means that it must not overlap with other patents, (iii) it must describe the invention and be in accordance with the feature described in the invention, (iv) each claim must stand on its own, and (v) each claim is written in a single sentenced and is numbered.

Independent claims are broad, which means that only the essential elements, in a generic form, are presented. In contrast, dependent claims are narrow by giving specifications of the components in a detailed manner to describe the preferred embodiment.

8.2.4.1 Grouping the claims

Normally, there is at least one main claim (independent claim) and a number of subsidiary claims (dependent claims), which are related to the former claim. The independent claim defines the invention by itself in a broader meaning. An independent claim does not require a limitation from another claim since it stands alone.

Usually, an independent claim is written including three statements: the preamble, the transition sentences, and the statements/details necessary for the invention to properly work. The transition sentences most employed are "consisting of" or "comprising." The independent claim contains elements of the invention, how these elements are connected, and how they play together for the functionality of the invention. For patents in chemistry, the independent claim usually starts with the main object of the invention, for example, "A polymeric matrix" and "A synthesis process." In these examples, the claim starts with the main problem that the present invention solves.

Patent application must have at least one independent claim. However, it is recommended having several independent claims, which individually covers different aspects of the invention. There is a balance between the number of claims and the costs of claim fees (annuity fees of a patent are based on the number of claims). The inventor needs to protect his or her invention by having a sufficient number of well-covered claims, without adding excessive costs. As stated before, claims are the heart of the invention, and therefore, they must be well written to avoid invalidation. Having well-writing claims, supported by the detailed description of the invention, makes the patent robust and difficult invalidation from the patent examiner or from a third party.

The dependent claim gives more details and explanation to the related independent claim. Independent claim is broader, while dependent claim narrows the related independent claim. This means that dependent claim must refer to an independent claim, by giving more details, new elements, new parameters, leading a limitation for the respectively independent claim. Caution must be taken since a dependent claim must not delete any element of the related independent claim. For example, in a chemical invention, an independent claim is based on the synthetic method to obtain a biomaterial. The corresponding dependent claim gives the details of the independent claim, for instance, the possible list of starting chemicals for the synthesis. A second dependent claim should give some experimental details and parameters in the synthesis to obtain the biomaterial described in claim 1, such as the addition of other chemicals or modifications of these chemical compounds. In this direction, a dependent claim refers back to an independent claim and incorporates or partly modifies the independent claim to create a narrower claim. It should be noted that claims are written in terms of specification, and the content of each claim must be presented and discussed in the detailed description of the embodiments or demonstrated in the drawings. The dependent claim should start with the following statement: "The chemical compound of Claim 1, further comprising ...." This statement clearly indicates that this is a dependent claim, and claim 1 is an independent claim. Usually, dependent claims are shorter than independent claims since the dependent claim is everything stated by the independent claims plus a limitation or a detail stated in the dependent claim. By definition, a dependent claim cannot extend the scope of invention protection stated by the associated independent claim. On contrary, a dependent claim limits the corresponding independent claim. However, care must be taken to avoid unnecessary limitation. In some cases, a dependent claim can refer to several dependent claims, according to the example: "The chemical product of claim 1 or 2, further modified …."

In chemical patent application, basically, there are two kinds of subject to be claimed: a process to obtain a product and/or a product. In the first case, claims should be written in a way that a skilled person in the art should be able to reproduce the process and obtain the product. In the case of claiming a product, the claim needs to clearly identify the product. Most of the claims based on products start from well-known products with some modifications, which improve the quality of the product, and are the subject of a patent. In this case, the product claim must clearly present the known product and the modifications. In other words, generally, there are two kinds of claim, in a chemical patent application. A claim can be based on a product (product claim), including the chemical composition of the compounds. A claim can also be based on a process to obtain the product, including the apparatus for the chemical reaction process. For example, "We claim a thiolated iron oxide nanoparticles …" (product claim). "We claim a method to thiolated iron oxide nanoparticles …" (process claim).

A claim is usually written in terms of positive features; however, in some cases, claims can be constructed in terms of negative features. For example, "A chemical reaction containing hydrophilic polymers, except polyethylene glycol." Such negative limitations might be used when adding positive features to the claim, not liming it. The aim of the negative claims is to remove nonpatentable features of the invention.

Some important tips in writing the claims are as follows:

- It is advised to first write broader claim (independent claim) and then write the corresponding limitations associated with the broader claim (dependent claims).
- Keep in mind that claims must be clear and objective. Therefore, avoid some adjectives such as "short," "fast," "perfect," and so on. Keep the text precise and accurate.
- Avoid ambiguity. The claims must be supported by the detailed description of the invention (embodiments) and/or drawings.
- Avoid inconsistency since it can invalidate the claim. For example, part of the detailed description of the invention and/or drawing is not covered by the claims. You need to remove this part from the detailed description of the invention and/or drawing or include a claim that covers this issue.
- Avoid *unnecessary* limitations of dependent claims. Revise the claims to check for unnecessary limitations.
- Claims must be written in terms of the present state of the art, which means that the claims must offer an advantage to the present state of art. The claim should overcome some limitations regarding the actual state of the art.

- In a chemical patent, if possible, write the claims in terms of product and process. Do not restrict the invention. On contrary, try to capture the full feature of the invention.
- Claim in chemistry might include distinct inventive parts: the compound itself, the modified compound with improved properties and/or utilities, other compositions containing the compound, the process for producing the compound, and the apparatus involved in the process. Try to claim all possible approaches.

8.2.5 *Drawings and brief description of the drawings*

Drawings are usually important to explain the invention allowing the disclosure. In some cases, including patents in chemistry, drawings are clear representation of the invention. As stated before, claims can be directly associated with drawings. In some cases, the drawings should support the invention, and in other cases, the drawings are an essential part of the patent application. In some cases, a schematic representation of the invention in the form of a drawing is the clearest form to describe an invention. Drawings should be prepared in a manner to sufficiently describe the invention (or part of the invention), and it must be in accordance with the patent text.

Each drawing must have a short description in words (the brief description of the drawings). For example, "Figure 1 is a schematic representation of the synthetic route to obtain target-site gold nanoparticles starting from ...." "Figure 2 is a diagrammatic representation of an embodiment in which the synthetic route is ...."

Important advice: drawings in a patent have a different concept from a scientific paper. In a patent, you have to spend skill and time to create original drawings to accurately describe the invention, not making them attractive. Usually, in patents, the drawings are made by simple sketch with a pencil on a paper. It is a different concept from scientific papers, in which the drawings are sometimes artistic, particularly in graphic abstract. Before elaborating your drawing for a patent, you should check the drawings in the patents.

8.2.6 *Background of the invention*

The key ideas in writing the background of the invention are as follows: (i) it should describe the actual state of the art, in which the invention is related. It is not the aim of the background of the invention to describe the invention, (ii) do not spend too much energy and time in writing the background. It should be short, usually 1–2 pages.

The background of the invention describes the actual state of the art, which means the technical field of the invention. The background section might not list the solutions. The solutions will be presented and discussed by the description of the invention. Thus, the background section merely describes the state of the art, independently of the invention.

8.2.7   Summary of the invention

The summary of the invention should highlight the important features of the invention, and it may have texts from the claims. It is the essence of the invention in a few sentences. Although in a patent the summary comes in the beginning of the text, it is recommended to write the summary after writing the detailed description of the invention and the claims. This strategy will allow the inventor to be able to summarize the invention in an accurate manner and in accordance with the content of the patent. It should be noted that not all jurisdictions require a summary of the invention. The summary section can start with the sentence: "A method to prepare gold nanoparticles …." Summary can describe the invention in very general terms. The summary section should finalize with a statement such as "These aspects and advantages of the present invention will become better understood upon further review of the claims and description sections of the invention."

8.2.8   Detailed description of the embodiments (examples)

This section is a comprehensive report of the description of the invention, including several examples (embodiments). The embodiments are variations of the invention. It should start from the broader description of the invention followed by the narrow/detailed description of the invention. In patents in chemistry, it should describe the chemical composition, the chemical compounds, the detailed process involved in the preparation of the material, the apparatus features, and so on. The detailed description of the invention must describe at least one embodiment. Usually, there are several embodiments. Each embodiment must be written with enough details, assuming that an ordinary skilled person would be able to follow it. The examples should be accurate and specific. Therefore, the key points in writing the detailed description of the invention are as follows:

- Start the description with a background information of the invention;
- Describe your invention in a broader manner;
- Describe your invention in a more detailed manner;
- It must contain at least one embodiment;
- The description text must be clear and accurate that someone skilled in your field would be able to repeat the invention by reading your instructions;
- Avoid inconsistency: The content of claims must be described in the detailed description of the invention.

8.3   Remarks

Experience will teach you to write a patent in an appropriate form. You should read other patents. This process is necessary since you have to know the actual state of the art in your field to demonstrate that you have an invention. Reading the actual state of the art includes a deep search in patents.

When writing a patent you should ask yourself the following:

- You should keep in mind: what are the goals of my invention?
- What must be protected in the claims?
- Who could license my patent?
- What are the advantages of this invention?
- How this invention might overcome the current limitations in the field that this patent belongs?
- Are the claims well written to efficiently protect the invention?
- Is the full text accurate and are the sections in accordance between them?
- Are the steps of the invention property described that someone skilled in the field would be able to reproduce the patent?

chapter nine

Final conclusions and perspectives

As discussed previously, a patent is a very specific right granted by the government of a country to an inventor, and this right gives the inventor to eliminate others from making, using, or selling this invention in that specific country during the patent's total life. It is important to know that a patent is issued to the individual inventor and not to a certain company; however, it is usually a typical; to have employees assign inventions to their employer. Remember that a patent protection is assessable for any product, procedure, or design that encounters certain demands of novelty, nonobviousness, and utility.

On the basis of all these data presented, in this book, we suggest an approach to read and construct a patent in chemistry and a strategy course for chemical students either at graduate or postgraduate level. However, introducing patent's concept into a core course for a chemistry institute or school requires an interdisciplinary outlook. Although our universities often expose the importance of interdisciplinary research and education in practice, academics and the courses they teach often remain locked in their own fields of specialization. We hope that this book will open the minds of the directors or rectors to understand that now it is time to review the way we teach our students. Finally, we expect that this book will help our students to visualize new strategies and new ways to enter into the entrepreneur class.

References

Barroso W, Quonian L, Pacheco E (2009) Patents as technological information in Latin America. *World Patent Inform* 31: 207–215.

Burn I (2014) How to read a patent. To understand a patent, it is essential to be able to read a patent. ATIPLaw (2014) (http://atintellectualproperty.com/how-to-read-patents) (Accessed in March 8, 2017).

CAGS (2015) A guide to intellectual property for graduate students and postdoctoral scholars. Canadian Association for Graduate Studies, Ottawa, Canada (www.cags.ca/documents/publications/working/Guide_Intellectual_Property.pdf).

Jansson U (2017) Patent documents as a source of technological information. WIPO roving national seminar on industrial property, Ethiopia (www.wipo.int/meetings/en/doc_details.jsp?doc_id=7557) (Accessed in March 8, 2017).

Jorge MF, Lopes FV, Barcelos VI, de Assis FLK, Travassos G, Freitas V, de Carvalho SP (2017) Bol Mens Propriedade Industrial, Rio de Janeiro, Brazil. 2: 1–18.

Kehoe C, Xiao JY (2001) Patent data for technology assessment, part I: Applications, patent databases, and retrieval methods. *Sci Technol Lib* 22(1/2): 101–116.

Mcleland L-N (2002) What every chemist should know about patents. ACS Joint Board–Council Committee on Patents and Related Matters (www.acs.org/content/dam/acsorg/about/governance/committees/what-every-chemist-should-know-about-patents.pdf) (Accessed in March 6, 2017).

Millman J (2014) Does it really cost $ 2.6 billion develop a new drug? *The Washington Post*, November 18.

Oubrich M, Barzi R (2014) Patents as a source of strategic information: The inventive activity in Morocco. *J Econ Int Bus Manage* 2: 27–35.

Sheldon JG (2016) How to write a patent application. Reading notes complied by David J. Stein ESq. third edition.

Shih MJ, Liu DR, Hsu ML (2010) Discovering competitive intelligence by mining changes in patent trends. *Expert Syst Appl* 37: 2882–2890.

van Overwalle G (2006) Intellectual property protection for medicinal and aromatic plants. In *Medicinal and Aromatic Plants* (R.J. Bogers, L.E. Craker and D. Lange (eds.)), Springer, The Netherlands, Chapter 9, pp. 121–128.

Walker B (2015) Innovation vs. invention: Make the leap and reap the rewards. *WIRE* (https://www.wired.com/insights/2015/01/innovation-vs-invention/) (Accessed in August 29, 2017).

Waller FJ (2011) *Writing Chemistry Patents and Intellectual Property: A Practical Guide*, John Wiley & Sons, Inc., Publ., NJ, USA, p. 256. ISBN 978-0-470-49740-1.

WIPO (2007a) Drafting patent manual (http://www.wipo.int/edocs/pubdocs/en/patents/867/wipo_pub_867.pdf) (Accessed in March 8, 2017).

WIPO (2007b) Patent drafting manual. World International Property Organization.

WIPO (2016) Global patent applications rose to 2.9 million in 2015 on strong growth from China; demand also increased for other intellectual property rights, Geneva, November (www.wipo.int/pressroom/en/articles/2016/article_0017.html) (Accessed in March 6, 2017).

Index